Design Recipes for FPGAs

Design Recipes for FPGAs

Dr Peter R. Wilson

AMSTERDAM • BOSTON • HEIDELBERG • LONDON • NEW YORK • OXFORD
PARIS • SAN DIEGO • SAN FRANCISCO • SINGAPORE • SYDNEY • TOKYO

Newnes is an imprint of Elsevier

Newnes is an imprint of Elsevier
Linacre House, Jordan Hill, Oxford OX2 8DP
30 Corporate Drive, Suite 400, Burlington MA 01803

First published 2007

British Library Cataloguing in Publication Data
Wilson, Peter R.
  Design recipes for FPGAs
  1. Field programmable gate arrays – Design and construction
  I. Title
  621.3′95

Library of Congress Number: 2007923611

ISBN: 978-0-7506-6845-3

For information on all Newnes publications
visit our website at www.books.elsevier.com

Printed and bound in Great Britain by MPG Books Ltd, Bodmin Cornwall

07 08 09 10 11   10 9 8 7 6 5 4 3 2 1

Cover image of an Actel RTAX4000S FPGA chip supplied courtesy of Actel – www.actel.com

For Heather

Contents

Acknowledgements

I would like to thank Professor Andrew Brown, the head of the Electronic Systems Design Group, School of Electronics and Computer Science, at the University of Southampton, UK. Giving me the opportunity to first study and then work in his group has led directly to me being able to write this book. For that I am deeply grateful. In addition, the continuing support and encouragement of colleagues and students in the ESD research group has been a constant source of support and ideas.

I also wish to single out Tim Pitts (Elsevier Publishing) who was instrumental in me starting this project, and also for his encouragement to see it through to a conclusion. I also would like to thank those who have contributed to the production of the book including Lisa Jones, Helen Eaton, Lewin Edwards, Charon Tec and team and all at Elsevier.

Finally a heartfelt thank you to all of my family, especially my wife Caroline, and children, Nathan and Heather. As always, without their support, none of this would be possible.

Peter R. Wilson

Preface

This book is designed to be a desktop reference for engineers, students and researchers who use Field Programmable Gate Arrays (FPGA) as their hardware platform of choice. This book has been produced in the spirit of the 'numerical recipe' series of books for various programing languages – where the intention is not to teach the language *per se*, but rather the philosophy and techniques required, making your application work. The rationale of this book is similar in that the intention is to provide the methods and understanding to make the reader able to develop practical, operational VHDL that will run correctly on FPGAs.

It is important to stress that his book is *not* designed as a language reference manual for VHDL. There are plenty of those available and I have referenced them throughout the text. This book is intended as a reference for design *with* VHDL and can be seen as complementary to a conventional VHDL textbook.

List of Figures

Part 1
Overview

The book is divided into five main parts. In the introductory part of the book, primers are given into Field Programmable Gate Arrays (FPGA), VHDL and the standard design flow. In the second part of the book, a series of complex applications that encompass many of the key design problems facing designers today are worked through from start to finish in a practical way. This will show how the designer can interpret a specification and develop a top-down design methodology and eventually build in detailed design blocks perhaps developed previously or by a third party. In the third part of the book, important techniques are discussed, worked through and explained from an example perspective, so you can see exactly how to implement a particular function. This part is really a toolbox of advanced specific functions that are commonly required in modern digital design. The fourth part on advanced techniques discusses the important aspect of design optimization, that is how can I make my design faster? Or more compact? The fifth part investigates the details of fundamental issues that are implemented in VHDL. This final part is aimed at designers with a limited VHDL background, perhaps those looking for simpler examples to get started, or to solve a particular detailed issue.

1

Introduction

Why FPGAs?

There are numerous options for designers in selecting a hardware platform for custom electronics design, ranging from embedded processors, Application Specific Integrated Circuits (ASICs), Programmable Micro-processors (PICs), FPGAs to Programmable Logic Devices (PLDs). The decision to choose a specific technology such as an FPGA should depend primarily on the design requirements rather than a personal preference for one technique over another.

For example, if the design requires a programmable device with many design changes, and algorithms using complex operations such asmultiplications and looping, then it may make more sense to use a dedicated signal processor device such as a DSP that can be programmed and reprogrammed easily using C or some other high-level language. If the speed requirements are not particularly stringent, and a compact cheap platform is required, then a general purpose microprocessor such as a PIC would be an ideal choice. Finally, if the hardware requirements require a higher level of performance, say up to several 100 MHz operation, then an FPGA offers a suitable level of performance, while still retaining the flexibility and reusability of programmable logic.

Other issues to consider are the level of optimization in the hardware design required. For example, a simple software program can be written in C, and then a PIC device programmed, but the performance may be limited by the inability of the processor to offer parallel operation of key functions. This can be implemented much more directly in an FPGA using parallelism and pipelining to achieve much greater throughput than would be possible using a PIC.

A general rule of thumb when choosing a hardware platform is to identify both the design requirements and the hardware options, and then select a suitable platform based on those considerations.

For example, if the design requires a basic clock speed of up to 100 MHz then an FPGA would be a suitable platform. If the clock speed could be 3–4 MHz, then the FPGA may be an expensive (overkill) option.

If the design requires a flexible processor option, although the FPGAs available today support embedded processors, it probably makes sense to use a DSP or PIC. If the design requires dedicated hardware functionality, then an FPGA is the route to take.

If the design requires specific hardware functions such as multiplication and addition, then a DSP may well be the best route, but if custom hardware design is required, then an FPGA would be the appropriate choice.

If the design requires small simple hardware blocks, then a PLD or CPLD (Complex Programmable Logic Device) may be the best option (compact, simple programmable logic), however, if the design has multiple functions, or a combination of complex controller and specific hardware functions, then the FPGA is the route to take.

Examples of this kind of decision can be dependent on the complexity of the hardware involved. For example, a Video Graphics Array (VGA) controller will probably require an FPGA rather than a PLD device, simply due to the complexity of the hardware involved. Another related issue is that of flexibility and programmability. If an FPGA is used, and the resources are not used upon a specific device (say up to 60 per cent for example), then if a communications protocol changes, or is updated, then the device may well have enough headroom to support several variants, or updates, in the future.

Using these simple guidelines, an intelligent choice can be made about the best platform to choose, and also which hardware device to select based on these assumptions. The nice aspect of most synthesis software packages is that multiple design platforms can be tested for performance and utilization (e.g. PLD or FPGA) prior to making a final decision on the hardware of choice.

2
An FPGA Primer

Introduction

This chapter is an introduction to the Field Programmable Gate Array (FPGA) platform for those unfamiliar with the technology. It is useful when designing hardware to understand the context that the hardware description language models (VHDL) are important and relevant to the ultimate design.

FPGA evolution

Since the inception of digital logic hardware in the 1970s, there has been a plethora of individual devices – leading to the ubiquitous TTL logic series still in use today (74/54 series logic), extended to CMOS technology (HC, AC, FC, FCT, HCT and so on). While these have been used extensively in Printed Circuit Board (PCB) design and still are today, there has been a consistent effort over the last 20 years to introduce greater programmability into basic digital devices.

The reason for this need is the dichotomy resulting from the two differing design approaches used for most digital systems. On the hardware side, the drive is usually toward ultimate performance: faster, smaller, lower power and cheaper. This leads to custom integrated circuit design (Application Specific Integrated Circuits or ASICs) where each chip (ASIC) has to be designed, laid out, fabricated and packaged individually. For large production runs this is cost effective, but obviously this approach is hugely expensive (masks alone for a current Silicon process may cost over $500 000) and time consuming (up to a year).

From a software perspective, however, the approach is more to use a standard processor architecture such as Intel Pentium, PowerPC or ARM, and develop software applications that can be downloaded onto such a platform. This type of approach is obviously quicker to implement a platform; however, usually there is a significant overhead due to the need for operating systems, compiler inefficiency and also a performance reduction due to the indirect relationship between the hardware and the software on the processor.

As a result, programmable devices have been developed as a form of intermediate approach; hardware design on a high-performance platform, optimal resources – no operating system required and reconfigurable as the devices can be reprogrammed.

Programmable logic devices

The first type of devices to be programmable was Programmable Array Logic (PAL). This consists of an array of logic gates that could be connected using an array of connections. These devices could support a small number of flip-flops (usually <10) and were able to implement small state machines (Figure 1).

Complex Programmable Logic Devices (CPLDs) were developed to address the limitations of simple PAL devices. These devices used the same basic principle as PALs, but had a series of macroblocks (each roughly equivalent to a PAL) and connected using routing blocks (Figure 2).

Field programmable gate arrays

The FPGAs were the next step from CPLD. Instead of a fixed array of gates, the FPGA uses the concept of a Complex Logic Block (CLB). This is configurable and allows not only routing on the device, but also each logic block can be configured optimally. A typical CLB is shown in Figure 3.

The CLB has a Look-Up Table (LUT) that can be configured to give a specific type of logic function when programmed. There is also a clocked d-type flip-flop that allows the CLB to be combinatorial (non-clocked) or synchronous (clocked), and there is also an enable signal. A Xilinx CLB is shown in Figure 4 and this shows

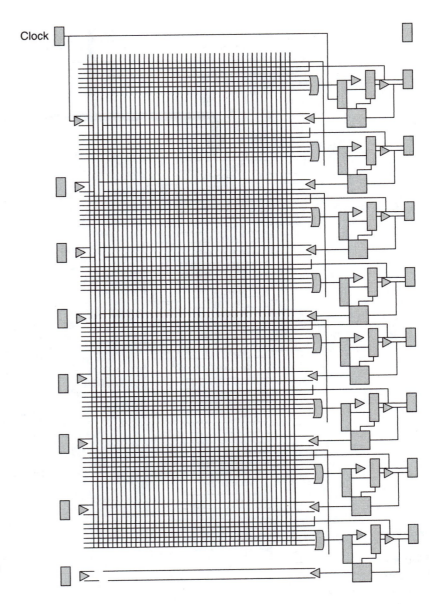

Clock

Figure 1
Programmable Logic
Device

clearly the two 4 input LUTs and various multiplexers and flip-flops in a real device.

A typical FPGA will have hundreds or thousands of CLBs, of different types, on a single device allowing very complex devices to be implemented on a single chip and configured easily. Modern FPGAs have enough capacity to hold a number of 32-bit processors on a single device. The layout of a typical FPGA (in CLB terms) is shown in Figure 5.

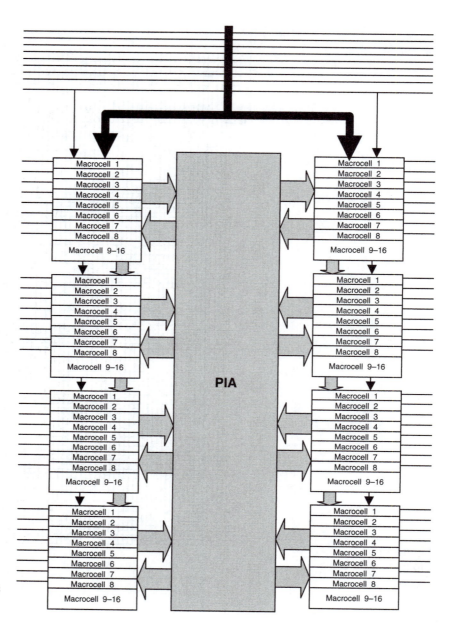

Figure 2
Complex
Programmable Logic
Device

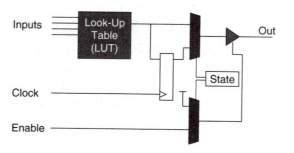

Figure 3
FPGA CLB

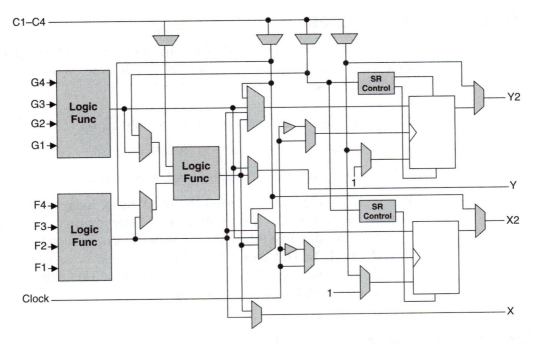

Figure 4
Xilinx CLB

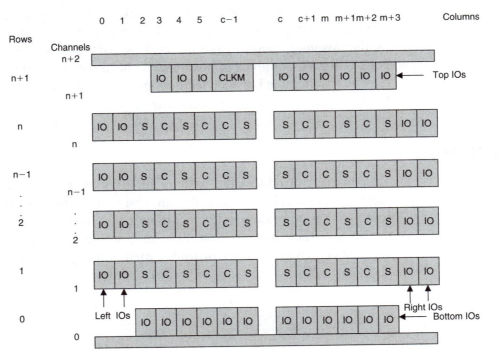

Figure 5
FPGA Structure of CLBs

FPGA design techniques

When we design using VHDL, these functions need to be mapped onto the low-level logic blocks on an FPGA. In order to do this, we need to carry out three specific functions:

1. *Mapping*: Logic functions mapped onto CLBs.
2. *Placement*: CLBs placed on FPGA.
3. *Routing*: Routed connections between CLBs.

It is clearly impossible to design 'by hand' using today's complex designs, we therefore rely on synthesis software to turn our VHDL design description into the logic functions that can be mapped onto the FPGA CLBs. This design flow is an iterative process including optimization and implies a complete design flow. This will be discussed in more detail later in this book.

Design constraints using FPGAs

It is very easy to produce unrealistic designs using VHDL if the target FPGA platform is not considered carefully. FPGAs obviously have a limited number of logic blocks and routing resources, and the design has to consider this. The style of VHDL code used by the designer should make the best use of resources, and this book will give examples of how that can be achieved. VHDL code may be transferable between technologies, but may need rewriting for best results due to these constraints.

Summary

This chapter introduces the basic technology behind FPGAs and their development. The key design issues are highlighted and some of the important design techniques introduced. Later chapters in this book will develop these in more detail either from a detailed design perspective or from a methodology point of view.

3

A VHDL Primer: The Essentials

Introduction

This chapter of the book is not intended as a comprehensive VHDL reference book – there are many excellent texts available that fit that purpose including Mark Zwolinski's *Digital System Design with VHDL*, Zainalabedin Navabi's VHDL: *Analysis and modeling of digital systems* or Peter Ashenden's *Designer's Guide to VHDL*. This section is designed to give concise and useful summary information on important language constructs and usage in VHDL – helpful and easy to use, but not necessarily complete.

This chapter will introduce the key concepts in VHDL and the important syntax required for most VHDL designs, particularly with reference to Field Programmable Gate Arrays (FPGAs). In most cases, the decision to use VHDL over other languages such as Verilog or SystemC, will have less to do with designer choice, and more to do with software availability and company decisions. Over the last decade or so, a 'war of words' has raged between the VHDL and Verilog communities about which is the best language, and in most cases it is completely pointless as the issue is more about design than syntax. There are numerous differences in the detail between VHDL and Verilog, but the fundamental philosophical difference historically has been the design context of the two languages. Verilog has come from a 'bottom-up' tradition and has been heavily used by the IC industry for cell-based design, whereas the VHDL language has been developed much more from a 'top-down' perspective. Of course, these are generalizations and largely out of date in a modern context, but the result is clearly seen in the basic syntax and methods of the two languages.

Without descending into a minute dissection of the differences between Verilog and VHDL one important advantage of VHDL is

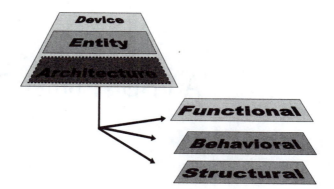

Figure 6
VHDL Models
with Different
Architectures

the ability to use multiple levels of model with different architectures as shown in Figure 6.

This is not unique to VHDL, and in fact Verilog does have the concept of different behavior in a single 'module'; however, it is explicitly defined in VHDL and is extremely useful in putting together practical multi-level designs in VHDL. The division of a model into its interface part (the 'entity' in VHDL) and the behavior part (the 'architecture' in VHDL) is an incredibly practical approach for modeling multiple behavior for a single interface and makes model exchange and multiple implementations straightforward.

The remainder of this chapter will describe the key parts of VHDL, starting with the definition of a basic model structure using entities and architectures, discuss the important variable types, review the methods of encapsulating concurrent, sequential and hierarchical behavior and finally introduce the important fundamental data types required in VHDL.

Entity: model interface

Entity definition

The entity defines how a design element described in VHDL connects to other VHDL models and also defines the name of the model. The entity also allows the definition of any parameters that are to be passed into the model using hierarchy. The basic template for an entity is as follows:

```
entity <name> is
....
entity <name>;
```

If the entity has the name 'test', then the entity template could be either:

```
entity test is
end entity test;
```

or:

```
entity test is
end test;
```

Ports

The method of connecting entities together is using PORTS. These are defined in the entity using the following method:

```
port (
...list of port declarations...
);
```

The port declaration defines the type of connection and direction where appropriate. For example, the port declaration for an input bit called in1 would be as follows:

```
in1 : in bit;
```

And if the model had two inputs (in1 and in2) of type bit and a single output (out1) of type bit, then the declaration of the ports would be as follows:

```
port (
        in1, in2 : in bit;
        out1 : out bit
);
```

As the connection points between entities are effectively the same as those inter-process connections, they are effectively signals and can be used as such within the VHDL of the model.

Generics

If the model has a parameter, then this is defined using generics. The general declaration of generics is shown below:

```
generic (
...list of generic declarations...
);
```

In the case of generics, the declaration is similar to that of a constant with the form as shown below:

```
param1 : integer := 4;
```

Taking an example of a model that had two generics (gain (integer) and time_delay (time)), they could be defined in the entity as follows:

```
generic (
      gain : integer := 4;
      time_delay : time = 10 ns
);
```

Constants

It is also possible to include model specific constants in the entity using the standard declaration of constants method previously described, for example:

```
constant : rpullup : real := 1000.0;
```

Entity examples

To illustrate a complete entity, we can bring together the ports and generics examples previously and construct the complete entity for this example:

```
entity test is
      port (
            in1, in2 : in bit;
            out1 : out bit
      );
      generic (
            gain : integer := 4;
            time_delay : time := 10 ns
      );
      constant : rpullup : real := 1000.0;
end entity test;
```

Architecture: model behavior

Basic definition of an architecture

While the entity describes the interface and parameter aspects of the model, the architecture defines the behavior. There are several types of VHDL architecture and VHDL allows different architectures to be defined for the same entity. This is ideal for developing behavioral, Register Transfer Level RTL and gate Level architectures that can be incorporated into designs and tested using the same test benches.

The basic approach for declaring an architecture could be as follows:

```
architecture behaviour of test is
..architecture declarations
begin
...architecture contents
end architecture behaviour;
```

or

```
architecture behaviour of test is
..architecture declarations
begin
...architecture contents
end behaviour;
```

Architecture declaration section

After the declaration of the architecture name and before the begin statement, any local signals or variables can be declared. For example, if there were two internal signals to the architecture called sig1 and sig2, they could be declared in the declaration section of the model as follows:

```
architecture behaviour of test is
      signal sig1, sig2 : bit;
begin
```

Then the signals can be used in the architecture statement section.

Architecture statement section

VHDL architectures can have a variety of structures to achieve different types of functionality. Simple combinatorial expressions use signal assignments to set new signal values as shown below:

```
out1 <= in1 and in2 after 10 ns;
```

Note that for practical design, the use of the 'after 10 ns' is not synthesizable. In practice, the only way to ensure correct synthesizable design is to either make the design delay insensitive or synchronous. The design of combinatorial VHDL will result is additional delays due to the technology library gate delays, potentially resulting in glitches or hazards. An example of a multiple gate combinatorial architecture using internal signal declarations is given below:

```
architecture behavioural of test is
      signal int1, int2 : bit;
begin
      int1 <= in1 and in2;
      int2 <= in3 or in4;
      out1 <= int1 xor int2;
end architecture behavioural;
```

Process: basic functional unit in VHDL

The process in VHDL is the mechanism by which sequential statements can be executed in the correct sequence, and with more than one process, concurrently. Each process consists of a sensitivity list, declarations and statements. The basic process syntax is given below:

```
process sensitivity_list is
    ... declaration part
begin
    ... statement part
end process;
```

The sensitivity list allows a process to be activated when a specific signal changes value, for example a typical usage would be to have a global clock and reset signal to control the activity of the process, for example:

```
process (clk, rst) is
begin
      ... process statements
end process;
```

In this example, the process would only be activated when either clk or rst changed value. Another way of encapsulating the same behavior is to use a wait statement in the process so that the process is automatically activated once, and then waits for activity on either signal before running the process again. The same process could then be written as follows:

```
process
begin
      ... process statements
      wait on clk, rst;
end process;
```

In fact, the location of the wait statement is not important, as the VHDL simulation cycle executes each process once during initialization, and so the wait statement could be at the start or the end of the process and the behavior would be the same in both cases.

In the declaration section of the process, signals and variables can be defined locally as described previously, for example a typical process may look like the following:

```
process (a) is
      signal na : bit;
begin
      na <= not a;
end process;
```

With the local signal na and the process activated by changes on the signal a that is externally declared (with respect to the process).

Basic variable types and operators

Constants

When a value needs to be static throughout a simulation, the type of element to use is a constant. This is often used to initialize parameters or to set fixed register values for comparison. A constant can be declared for any defined type in VHDL with examples as follows:

```
constant a : integer := 1;
constant b : real := 0.123;
constant c : std_logic := '0';
```

Signals

Signals are the link between processes and sequential elements within the processes. They are effectively 'wires' in the design and connect all the design elements together. When simulating signals, the simulator will in turn look at updating the signal values and also checking the sensitivity lists in processes to see whether any changes have occurred that will mean that processes become active.

Signals can be assigned immediately or with a time delay, so that an event is scheduled for sometime in the future (after the specified delay). It is also important to recognize that signals are not the same as a set of sequential program code (such as in C), but are effectively concurrent signals that will not be able to be considered stable until the next time the process is activated.

Examples of signal declaration and assignment are shown below:

```
signal sig1 : integer := 0;
signal sig2 : integer := 1;
sig1 <= 14;
sig1 <= sig2;
sig1 <= sig2 after 10 ns;
```

Variables

While signals are the external connections between processes, variables are the internal values within a process. They are only used in a sequential manner, unlike the concurrent nature of signals within and between processes. Variables are used within processes and are declared and used as follows:

```
variable var1 : integer := 0;
variable var2 : integer := 1;
var1 := var2;
```

Notice that there is no concept of a delay in the variable assignment – if you need to schedule an event, it is necessary to use a signal.

Boolean operators

VHDL has a set of standard Boolean operators built in, which are self-explanatory. The list of operators are and, or, nand, not, nor, xor. These operators can be applied to BIT, BOOLEAN or logic types with examples as follows:

```
out1 <= in1 and in2;
out2 <= in3 or in4;
out5 <= not in5;
```

Arithmetic operators

There are a set of arithmetic operators built into VHDL which again are self-explanatory and these are described and examples provided, see next page.

Operator	Description	Example
+	Addition	out1 <= in1 + in2;
−	Subtraction	out1 <= in1 − in2;
*	Multiplication	out1 <= in1 * in2;
/	Division	out1 <= in1/in2;
abs	Absolute Value	absin1 <= abs(in1);
mod	Modulus	modin1 <= mod(in1);
rem	Remainder	remin1 <= rem(in1);
**	Exponent	out1 <= in1 ** 3;

Comparison operators

VHDL has a set of standard comparison operators built in, which are self-explanatory. The list of operators are =, /=, <, <=, >, >=. These operators can be applied to a variety of types as follows:

```
in1 < 1
in1 /= in2
in2 >= 0.4
```

Shifting functions

VHDL has a set of six built in logical shift functions which are summarized below:

Operator	Description	Example
sll	Shift Left Logical	reg <= reg sll 2;
srl	Shift Right Logical	reg <= reg srl 2;
sla	Shift Left Arithmetic	reg <= reg sla 2;
sra	Shift Right Arithmetic	reg <= reg sra 2;
rol	Rotate Left	reg <= reg rol 2;
ror	Rotate Right	reg <= reg ror 2;

Concatenation

The concatenation function XE 'VHDL:concatenation' in VHDL is denoted by the & symbol and is used as follows:

```
A <= '1111';
B <= '000';
out1 <= A & B & '1'; -- out1 = '11110001';
```

Decisions and loops

If-then-else

The basic syntax for a simple if statement is as follows:

```
if (condition) then
        ... statements
end if;
```

The condition is a Boolean expression, of the form a > b or
a = b. Note that the comparison operator for equality is a single
= , not to be confused with the double == used in some programming languages. For example, if two signals are equal, then set an
output high would be written in VHDL as:

```
if ( a = b ) then
        out1 <= '1';
end if;
```

If the decision needs to have both the if and else options, then the
statement is extended as follows:

```
if (condition) then
        ... statements
else
        ... statements
end if;
```

So in the previous example, we could add the else statements as
follows:

```
if ( a = b ) then
        out1 <= '1';
else
        out1 <= '0';
end if;
```

And finally, multiple if conditions can be implemented using the
general form:

```
if (condition1) then
        ... statements
elsif (condition2)
        ... statements
... more elsif conditions & statements
else
        ... statements
end if;
```

With an example:

```
if (a > 10) then
      out1 <= '1';
elsif (a > 5) then
      out1 <= '0';
else
      out1 <= '1';
end if;
```

Case

As we have seen with the IF statement, it is relatively simple to define multiple conditions, but it becomes a little cumbersome, and so the case statement offers a simple approach to branching, without having to use Boolean conditions in every case. This is especially useful for defining state diagrams or for specific transitions between states using enumerated types. An example of a case statement is:

```
case testvariable is
      when 1 =>
              out1 <= '1';
      when 2 =>
              out2 <= '1';
      when 3 =>
              out3 <= '1';
end case;
```

This can be extended to a range of values, not just a single value :

```
case test is
      when 0 to 4 => out1 <= '1';
```

It is also possible to use Boolean conditions and equations. In the case of the default option (i.e. when none of the conditions have been met), then the term when others can be used:

```
case test is
      when 0 => out1 <= '1';
      when others => out1 <= '0';
end case;
```

For

The most basic loop in VHDL is the FOR loop. This is a loop that executes a fixed number of times. The basic syntax for the FOR loop is shown below:

```
for loopvar in start to finish loop
      ... loop statements
end loop;
```

It is also possible to execute a loop that counts down rather than up, and the general form of this loop is:

```
for loopvar in start downto finish loop
     ... loop statements
end loop;
```

A typical example of a for loop would be to pack an array with values bit by bit, for example:

```
signal a : std_logic_vector(7 downto 0);
for i in 0 to 7 loop
     a(i) <= '1';
end loop;
```

While and loop

Both the while and loop loops have an in-determinant number of loops, compared to the fixed number of loops in a FOR loop and as such are usually not able to be synthesized. For FPGA design, they are not feasible as they will usually cause an error when the VHDL model is compiled by the synthesis software.

Exit

The exit command allows a FOR loop to be exited completely. This can be useful when a condition is reached and the remainder of the loop is no longer required. The syntax for the exit command is shown below:

```
for i in 0 to 7 loop
     if ( i = 4 ) then
            exit;
     endif;
endloop;
```

Next

The next command allows a FOR loop iteration to be exited, this is slightly different to the exit command in that the current iteration is exited, but the overall loop continues onto the next iteration. This can be useful when a condition is reached and the remainder of the iteration is no longer required. An example for the next command is shown below:

```
for i in 0 to 7 loop
     if ( i = 4 ) then
            next;
     endif;
endloop;
```

Hierarchical design

Functions

Functions are a simple way of encapsulating behavior in a model that can be reused in multiple architectures. Functions can be defined locally to an architecture or more commonly in a package (discussed in the next section of this book), but in this section the basic approach of defining functions will be described. The simple form of a function is to define a header with the input and output variables as shown below:

```
function name (input declarations) return output_type is
      ... variable declarations
begin
      ... function body
end
```

For example, a simple function that takes two input numbers and multiplies them together could be defined as follows:

```
function mult (a,b : integer) return integer is
begin
      return a * b;
end;
```

Packages

Packages are a common single way of disseminating type and function information in the VHDL design community. The basic definition of a package is as follows:

```
package name is
...package header contents
end package;
package body name is
      ... package body contents
end package body;
```

As can be seen, the package consists of two parts, the header and the body. The header is the place where the types and functions are declared, and the package body is where the declarations themselves take place.

For example, a function could be described in the package body and the function is declared in the package header. Take a simple example of a function used to carry out a simple logic function:

```
and10 = and(a,b,c,d,e,f,g,h,i,j)
```

The VHDL function would be something like the following:

```
function and10 (a,b,c,d,e,f,g,h,i,j : bit) return bit is
begin
      return a and b and c and d and e and f and g and h
        and i and j;
end;
```

The resulting package declaration would then use the function in the body and the function header in the package header thus:

```
package new_functions is
function and10 (a,b,c,d,e,f,g,h,i,j : bit) return bit;
end;
package body new_functions is
     function and10 (a,b,c,d,e,f,g,h,i,j : bit) return
       bit is
    begin
          return a and b and c and d and e \
            and f and g and h and i and j;
    end;
end;
```

Components

While procedures, functions and packages are useful in including behavioral constructs generally, with VHDL being used in a hardware design context, often there is a need to encapsulate design blocks as a separate component that can be included in a design, usually higher in the system hierarchy. The method for doing this in VHDL is called a COMPONENT. Caution needs to be exercised with components as the method of including components changed radically between VHDL 1987 and VHDL 1993, as such care needs to be taken to ensure that the correct language definitions are used consistently.

Components are a way of incorporating an existing VHDL entity and architecture into a new design without including the previously created model. The first step is to declare the component – in a similar way that functions need to be declared. For example, if an entity is called and4, and it has 4 inputs (a, b, c, d of type bit) and 1 output (q of type bit), then the component declaration would be of the form shown below:

```
component and4
     port ( a, b, c, d : in bit; q : out bit );
end component;
```

Then this component can be instantiated in a netlist form in the VHDL model architecture:

```
d1 : and4 port map ( a, b, c, d, q );
```

Note that in this case, there is no explicit mapping between port names and the signals in the current level of VHDL, the pins are mapped in the same order as defined in the component declaration. If each pin is to be defined independent of the order of the pins, then the explicit port map definition needs to be used:

```
d1: and4 port map ( a => a, b => b, c => c, d => d, q =>
   q);
```

The final thing to note is that this is called the default binding. The binding is the link between the compiled architecture in the current library and the component being used. It is possible, for example, to use different architectures for different instantiated components using the following statement for a single specific device:

```
for d1 : and4 use entity work.and4(behaviour) port map
   (a,b,c,d,q);
```

or the following to specify a specific device for all the instantiated components:

```
for all : and4 use entity work.and4(behaviour) port
   map (a,b,c,d,q);
```

Procedures

Procedures are similar to functions, except that they have more flexibility in the parameters, in that the direction can be in, out or inout. This is useful in comparison to functions where there is generally only a single output (although it may be an array) and avoids the need to create a record structure to manage the return value. Although procedures are useful, they should be used only for small specific functions. Components should be used to partition the design, not procedures, and this is especially true in FPGA design, as the injudicious use of procedures can lead to bloated and inefficient implementations, although the VHDL description can be very compact. A simple procedure to execute a full adder could be of the form:

```
procedure full_adder (a,b : in bit; sum, carry : out bit)
   is
begin
      sum := a xor b;
      carry := a and b;
   end;
```

Notice that the syntax is the same as that for variables (NOT signals), and that multiple outputs are defined without the need for a return statement.

Debugging models

Assertions

Assertions are used to check if certain conditions have been met in the model and are extremely useful in debugging models. Some examples:

```
assert value <= max_value
       report "Value too large";
assert clock_width >= 100 ns
       report "clock width too small"
       severity failure;
```

Basic data types

Basic types

VHDL has the following standard types defined as built in data types:

- BIT
- BOOLEAN
- BIT_VECTOR
- INTEGER
- REAL

Data type: BIT

The BIT data type is the simple logic type built into VHDL. The type can have two legal values '0' or '1'. The elements defined as of type BIT can have the standard VHDL built in logic functions applied to them. Examples of signal and variable declarations of type BIT follow:

```
signal ina : bit;
variable inb : bit := '0';
ina <= inb and inc;
ind <= '1' after 10 ns;
```

Data type: Boolean

The Boolean data type is primarily used for decision-making, so the test value for 'if' statements is a Boolean type. The elements defined as of type Boolean can have the standard VHDL built in logic functions applied to them. Examples of signal and variable declarations of type Boolean follow:

```
signal test1 : Boolean;
variable test2 : Boolean := FALSE;
```

Data type: integer

The basic numeric type in VHDL is the integer and is defined as an integer in the range -2147483647 to $+2147483647$. There are obviously implications for synthesis in the definition of integers in any VHDL model, particularly the effective number of bits, and so it is quite common to use a specified range of integer to constrain the values of the signals or variables to within physical bounds. Examples of integer usage follow:

```
signal int1 : integer;
variable int2 : integer := 124;
```

There are two subtypes (new types based on the fundamental type) derived from the integer type which are integer in nature, but simply define a different range of values.

Integer subtypes: natural

The natural subtype is used to define all integers greater than and equal to zero. They are actually defined with respect to the high value of the integer range as follows:

```
natural values : 0 to integer'high
```

Integer subtypes: positive

The positive subtype is used to define all integers greater than and equal to one. They are actually defined with respect to the high value of the integer range as follows:

```
positive values : 1 to integer'high
```

Data type: character

In addition to the numeric types inherent in VHDL, there are also the complete set of ASCII characters available for designers. There is no automatic conversion between characters and a numeric value *per se*; however, there is an implied ordering of the characters defined in the VHDL standard (IEEE Std 1076-1993). The characters can be defined as individual characters or arrays of characters to create strings. The best way to consider characters is an enumerated type.

Data type: real

Floating point numbers are used in VHDL to define real numbers and the predefined floating point type in VHDL is called real. This defines a floating point number in the range $-1.0e38$ to $+10e38$. This is an important issue for many FPGA designs, as most commercial synthesis products do not support real numbers – precisely because they are floating point. In practice, it is necessary to use integer or fixed point numbers which can be directly and simply synthesized into hardware. An example of defining real signals or variables is shown below:

```
signal realno : real;
variable realno : real := 123.456;
```

Data type: time

Time values are defined using the special time type. These not only include the time value, but also the unit – separated by a space. The basic range of the time type value is between -2147483647 and 2147483647, and the basic unit of time is defined as the femto-second (fs). Each subsequent time unit is derived from this basic unit of the fs as shown below:

```
ps  = 1000 fs;
ns  = 1000 ps;
us  = 1000 ns;
ms  = 1000 us;
min = 60 sec;
hr  = 60 min;
```

Examples of time definitions are shown below:

```
delay : time := 10 ns;
wait for 20 us;
y <= x after 10 ms;
z <= y after delay;
```

Summary

This chapter provides a very brief introduction to VHDL and is certainly not a comprehensive reference. It enables the reader, hopefully, to have enough knowledge to understand the syntax of the examples in this book. The author strongly recommends that anyone serious about design with VHDL should also obtain a detailed and comprehensive reference book on VHDL, such as Zwolinski (a useful introduction to digital design with VHDL – a common student textbook) or Ashenden (a more heavy duty VHDL reference that is perhaps more comprehensive, but less easy for a beginner to VHDL).

4

Design Automation and Testing for FPGAs

Simulation

Test benches

The overall goal of any hardware design is to ensure that the design meets the requirements of the design specification. In order to measure this is indeed the case we need to not only simulate the design representation in a hardware description language (such as VHDL), but also to ensure that whatever tests we undertake are appropriate and demonstrate that the specification has been met.

The way that designers can test their designs in a simulator is by creating a 'test bench'. This is directly analogous to a real experimental test bench in the sense that stimuli are defined, and the responses of the circuit measured to ensure that they meet the specification.

In practice, the test bench is simply a VHDL model that generates the required stimuli and checks the responses. This can be in such a way that the designer can view the waveforms and manually check them, or by using VHDL constructs to check the design responses automatically.

Test bench goals

The goals of any test bench are twofold. The first is primarily to ensure that correct operation is achieved. This is essentially a 'functional' test. The second goal is to ensure that a synthesised design still meets the specification (particularly with a view to timing errors).

Simple test bench: instantiating components

Consider a simple combinatorial VHDL model given below:

```
library ieee;
use ieee.std_logic_1164.all;
entity cct is
        port (    in0, in1 : in std_logic;
                    out1 : out std_logic
        );
end;

architecture simple of cct is
begin
        out1 <= in0 AND in1 ;
end;
```

This simple model is clearly a two input AND gate, and to test the operation of the component we need to do several things.

First, we must include the component in a new VHDL design. So we need to create a basic test bench. The listing below shows how a basic entity (with no connections) is created, and then the architecture contains both the component declaration and the signals to test the design.

```
-- library declarations
library ieee;
use ieee.std_logic_1164.all;

-- empty entity declaration
entity test is
end;

-- test bench architecture
architecture testbench of test is
        -- component declaration
        component cct
            port (  in0, in1 : in std_logic;
                    out1 : out std_logic
            );
        end component;
        -- test bench signal declarations
        signal in0, in1, out1 : std_logic;
-- architecture body
Begin
        -- declare the Circuit Under Test (CUT)
        CUT : cct port map ( in0, in1, out1 );
end;
```

This test bench will compile in a VHDL simulator, but is not particularly useful as there are no definitions of the input stimuli (signals in0 and in1) that will exercise the Circuit Under Test (CUT).

If we wish to add stimuli to our test bench we have some significant advantages over our design VHDL – the most appealing is that we generally don't need to adhere to any design rules or even make the code synthesisable. Test bench code is generally designed to be 'off chip' and therefore we can make the code as abstract or behavioral as we like and it will still be fit for purpose. We can use wait statements, file read and write, assertions and other non-synthesisable code options.

Adding stimuli

In order to add a basic set of stimuli to our test bench, we could simply define the values of the input signals in0 and in1 with a simple signal assignment:

```
begin
      CUT : cct port map ( in0, in1, out1 );

      in0 <= '0';
      in1 <= '1';
end;
```

Clearly this is not very complex or dynamic test bench, so to add a sequence of events we can modify the signal assignments to include numerous value, time pairs defining a sequence of values.

```
begin
      CUT : cct port map ( in0, in1, out1 );

      in0 <= '0' after 0 ns, '1' after 10 ns, '0' after
         20 ns;
      in1 <= '0' after 0 ns, '1' after 15 ns, '0' after
         25 ns;
end;
```

While this method is useful for small circuits, clearly for more complex realistic designs it is of limited value. Another approach is to define a constant array of values that allow a number of tests to be carried out with a relatively simple test bench and applying a different set of stimuli and responses in turn.

For example, we can exhaustively test our simple two input logic design using a set of data in a **record**. A VHDL record is simply a collection of types grouped together defined as a new type.

```
type testdata is record
      in0 : std_logic;
      in1 : std_logic;
end;
```

With a new composite type, such as a record, we can then create an array, just as in any standard VHDL type. This requires another type declaration, of the array type itself.

```
type data_array is array (natural range <>) of data_array
```

With these two new types we can simply declare a constant (of type data_array) that is an array of record values (of type testdata) that fully describe the data set to be used to test the design. Notice that the type data_array does not have a default range, but that this is defined by the declaration in this particular test bench.

```
constant test_data : data_array := ( ('0', '0'), ('0',
 '1'), ('1', '0'), ('1', '1') );
```

The beauty of this approach is that we can change from a system that requires every test stimulus to be defined explicitly, to one where a generic test data process will read values from pre-defined arrays of data. In the simple test example presented here, an example process to apply each set of test data in turn could be implemented as follows:

```
process
begin
        for i in test_data'range loop
                in0 <= test_data(i).in0;
                in1 <= test_data(i).in1;
                wait for 100 ns;
        end loop
        wait;
end process;
```

There are several interesting aspects to this piece of test bench VHDL. The first is that we can use behavioral VHDL (wait for 100 ns) as we are not constrained to synthesize this to hardware. Secondly, by using the range operator, the test bench becomes unconstrained by the size of the data set. Finally, the individual record elements are accessed using the hierarchical construct test_data(i).in0 or test_data(i).in1, respectively.

Libraries

Introduction

VHDL as a language on its own is actually very limited in the breadth of the data types and primitive models available. As a result, libraries are required to facilitate design re-use and standard data types for model exchange, re-use and synthesis. The primary library for standard VHDL design is the IEEE library. Within the IEEE Design Automation Standards Committee (DASC), various committees have developed libraries, packages and extensions to standard VHDL. Some of these are listed below:

- IEEE Std 1076 Standard VHDL Language

- IEEE Std 1076.1 Standard VHDL Analog and Mixed-Signal Extensions (VHDL-AMS)

- IEEE Std 1076.1.1 Standard VHDL Analog and Mixed-Signal Extensions – Packages for Multiple Energy Domain Support

- IEEE Std 1076.4 Standard VITAL ASIC (Application Specific Integrated Circuit) Modeling Specification (VITAL)

- IEEE Std 1076.6 Standard for VHDL Register Transfer Level (RTL) Synthesis (SIWG)

- IEEE Std 1076.2 IEEE Standard VHDL Mathematical Packages (math)

- IEEE Std 1076.3 Standard VHDL Synthesis Packages (vhdlsynth)

- IEEE Std 1164 Standard Multivalue Logic System for VHDL Model Interoperability (Std_logic_1164)

Each of these 'working groups' are volunteers who come from a combination of academia, EDA industry and user communities, and collaborate to produce the IEEE Standards (usually revised every 4 years).

Using libraries

In order to use a library, first the library must be declared:

```
library ieee;
```

Within each library a number of VHDL packages are defined, that allow specific data types or functions to be employed in the design. For example, in digital systems design, we require logic data types, and these are not defined in the basic VHDL standard (1076). Standard VHDL defines integer, boolean and bit types, but not a standard logic definition. This is obviously required for digital design and an appropriate IEEE standard was developed for this purpose – IEEE 1164. It is important to note that IEEE Std 1164 is

NOT a subset of VHDL (IEEE 1076), but is defined for hardware description languages in general.

Std_logic libraries

There are a number of std_logic libraries available in the IEEE library and these are:

- std_logic_1164
- std_logic_arith
- std_logic_unsigned
- std_logic_signed
- std_logic_entities
- std_logic_components
- std_logic_misc
- std_logic_textio

In order to use a particular element of a package in a design, the user is required to declare their use of a package using the USE command. For example, to use the standard IEEE logic library, the use needs to add a declaration after the library declaration as follows:

```
library ieee;
use ieee.std_logic_1164.all;
```

The std_logic_1164 package is particularly important for most digital design, especially for Field Programmable Gate Array (FPGA), because it defines the standard logic types used by ALL the commercially available simulation and synthesis software tools, and is included as a standard library. It incorporates not only the definition of the standard logic types, but also conversion functions (to and from the standard logic types) and also manages the conversion between signed, unsigned and logic array variables.

Std_logic type definition

As it is such an important type, the std_logic type is described in this section. The type has the following definition:

- 'U': uninitialized; this signal hasn't been set yet
- 'X': unknown; impossible to determine this value/result
- '0': logic 0

- '1': logic 1
- 'Z': High Impedance
- 'W': Weak signal, can't tell if it should be 0 or 1
- 'L': Weak signal that should probably go to 0
- 'H': Weak signal that should probably go to 1
- '-': Don't care

These definitions allow resolution of logic signals in digital designs in a standard manner that is predictable and repeatable across software tools and platforms. The operations that can be carried out on the basic std_logic data types are the standard built in VHDL logic functions:

- and
- nand
- or
- nor
- xor
- xnor
- not

An example of the use of the std_logic library would be to define a simple logic gate – in this case a three input nand gate:

```
library ieee;
use ieee.std_logic_1164.all;

entity nand3 is
     port ( in0, in1, in2 : in std_logic;
                  out1 : out std_logic ) ;
end;

architecture simple of nand3 is
begin
     out1 <= in0 nand in1 nand in2;
end;
```

Synthesis

Design flow for synthesis

The basic HDL design flow is shown in Figure 7.

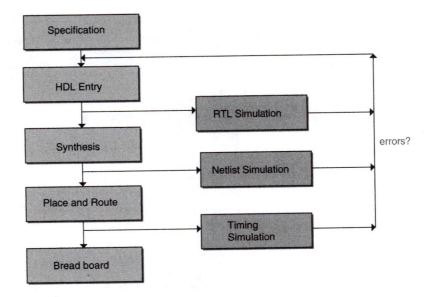

Figure 7
HDL Design Flow

As can be seen from this figure, Synthesis is the key stage between high-level design and the physical place and route which is the final product of the design flow. There are several different types of synthesis ranging from behavioral, to RTL and finally physical synthesis.

Behavioral synthesis is the mechanism by which high level abstract models are synthesized to an intermediate model that is physically realizable. Behavioral models can be written in VHDL that are not directly synthesizable and so care must be taken with high level models to ensure that this can take place, in fact. There are limited tools that can synthesize behavioral VHDL and these include the Behavioral Compiler from Synopsys, Inc and MOODS, a research synthesis platform from the University of Southampton.

RTL Synthesis is what most designers call synthesis, and is the mechanism whereby a direct translation of structural and register level VHDL can be synthesized to individual gates targeted at a specific FPGA platform. At this stage, detailed timing analysis can be carried out and an estimate of power consumption obtained. There are numerous commercial synthesis software packages including Design Compiler (Synopsys), Leonardo Spectrum (Mentor Graphics) and Synplify (Synplicity) – but this is not an exhaustive list – there are numerous offerings available at a variety of prices.

Physical synthesis is the last stage in a synthesis design flow and is where the individual gates are placed (using a 'floorplan') and routed on the specific FPGA platform.

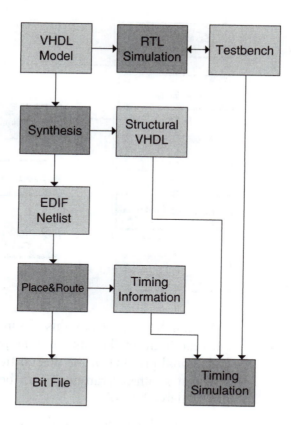

Figure 8
RTL Synthesis and
Design Flow

Synthesis issues

Synthesis basically transforms program-like VHDL into a true hardware design (netlist). It requires a set of inputs, a VHDL description, timing constraints (when outputs need to be ready, when inputs will be ready, data to estimate wire delay), a technology to map to (list of available blocks and their size/timing information) and information about design priorities (area vs. speed)

For big designs, the VHDL will typically be broken into modules and then synthesized separately. 10 K gates per module was a reasonable size in the 1990s, however tools can handle a lot more now.

RTL design flow

RTL VHDL is the input to most standard synthesis software tools. The VHDL must be written in a form that contains Registers, State Machines (FSM) and combinational logic functions. The synthesis software translates these blocks and functions into gates and library cells from the FPGA library. The RTL design flow is shown in Figure 8, in more detail than the overall HDL design flow. Using

RTL VHDL restricts the scope of the designer as it precludes algorithmic design – as we shall see later. This approach forces the designer to think at quite a low level – making the resulting code sometimes verbose and cumbersome. It also forces structural decisions early in the design process – restrictive and not always advisable, or helpful.

The Design process starts from RTL VHDL:

- Simulation (RTL) – is needed to develop a test bench (VHDL).

- Synthesis (RTL) – targeted at a standard FPGA platform.

- Timing simulation (Structural) – simulate to check timing.

- Place and route using standard tools (e.g. Xilinx Design Manager).

Although there are a variety of software tools available for synthesis (such as Leonardo Spectrum or Synplify), they all have generally similar approaches and design flows.

Physical design flow

Synthesis generates a netlist of devices plus interconnections. The 'place and route' software figures out where the devices go and how to connect them. The results not as good as you'd like; a 40 to 60 per cent utilization of devices and wires is typical. The designer can trade off run time against greater utilization to some degree, but there are serious limits. Typically the FPGA vendor will provide a software toolkit (such as the Xilinx Design Navigator or Altera's Quartus tools) that manages the steps involved in physical design.

Regardless of the particular physical synthesis flow chosen, the steps required to translate the VHDL or EDIF output from an RTL synthesis software program into a physically downloadable bit file are essentially the same and are listed below:

1. Translate

2. Map

3. Place

4. Route

5. Generate accurate timing models and reports

6. Create binary files for download to device

Place and route

There are two main techniques to place and route in current commercial software which are recursive cut and simulated annealing.

Recursive cut

In a recursive cut algorithm, we divide the netlist into two halves, move devices between halves to minimize the number of wires that cross cut (while keeping the number of devices in each half the same). This is repeated to get smaller and smaller blocks.

Timing analysis

Static timing analysis is the most commonly-used approach. In static timing analysis, we calculate the delay from each input to each output of all devices. The delays are added up along each path through circuit to get the critical path through the design and hence the fastest design speed.

This works as long as there are no cycles in the circuit, however in these cases the analysis becomes less easy. Design software allows you to break cycles at registers to handle feedback if this is the case.

As in any timing analysis, the designer can trade off some accuracy for run time. Digital simulation software such as Modelism or Verilog will give fast results, but will use approximate models of timing, whereas analog simulation tools like SPICE will give more accurate numbers, but take much longer to run.

Design pitfalls

The most common mistake that inexperienced designers make is simply making things too complex. The best approach to successful design is to keep the design elements simple, and the easiest way to manage that is efficient use of hierarchy.

The second mistake that is closely related to design complexity is not testing enough. It is vital to ensure that all aspects of the design are adequately tested. This means not only carrying out basic functional testing, but also systematic testing, and checking for redundant states and potential error states.

Another common pitfall is to use multiple clocks unnecessarily. Multiple clocks can create timing related bugs that are transient or hardware dependent. They can also occur in hardware and yet be missed by simulation.

VHDL issues for FPGA design

Initialization

Any default values of signals and variables are ignored. This means that you must ensure that synchronous (or asynchronous) sets and resets must be used on all flip-flops to ensure a stable starting condition. Remember that synthesis tools are basically stupid and follows a basic set of rules that may not always result in the hardware that you expect.

Floating point numbers and operations

Data types using floating point are currently not supported by synthesis software tools. They generally require 32 bits and the requisite hardware is just too large for most FPGA and ASIC platforms.

Summary

This chapter has introduced the practical aspect of developing test benches and validating VHDL models using simulation. This is an often overlooked skill in VHDL (or any hardware description language) and is vital to ensuring correct behavior of the final implemented design. We have also introduced the concept of design synthesis and highlighted the problem of not only ensuring that a design simulates correctly, but also how we can make sure that the design will synthesize to the target technology and still operate correctly with practical delays and parasitics. Finally, we have raised some of the practical implementation issues and potential problems that can occur with real designs, and these will be discussed in more detail in Part 4 of this book.

An important concept useful to define here is the difference between validation and verification. The terms are often confused leading to problems in the final design and meeting a specification. Validation is the task of ensuring that the design is 'doing the right thing'. If the specification asks for a low pass filter, then we must implement a low pass filter to have a valid design. We can even be more specific and state that the design must perform within a constraint. Verification, on the other hand, is much more specific and can be stated as 'doing the right thing *right*'. In other words, verification is ensuring that not only does our design do what is required functionally, but in addition it must meet ALL the criteria defined by the specification, preferably with some head-room to ensure that the design will operate to the specification under all possible operating conditions.

Part 2
Applications

The aim is of the applications section of this book is to identify key points/issues and 'nuggets' of information that are of practical use to the designer. The technical information on the issues provided later in the book are referenced, enabling the reader to see the 'wood for the trees' and select the 'trees' they need to solve a particular issues. Each application uses a combination of block diagrams, state diagrams and code snippets to explain the key concepts in making the application work. Detailed analysis of specific aspects of the design are forward referenced as required.

The first application is a high speed video monitor system that requires the implementation of a link to a video camera, and also interfaces to Random Access Memory (RAM) and a hard disk. While this is a notional system, the concept is in common usage in a variety of industries. The techniques involved are in exactly the area that is useful for a wide range of similar testing and monitoring applications.

The second application is more about processing power and illustrates the practical aspects of developing multiple processor cores on a standard Field Programmable Gate Array (FPGA) platform and how that can be managed in practice.

5

Images and High-Speed Processing

Introduction

This application is designed to show how several high data rate applications can be handled using VHDL on FPGAs (Field Programmable Gate Arrays). The system consists of a high speed camera, processor core, disk drive interface, Random Access Memory (RAM) interface and serial link to an external Program Counter (PC). The overall system has been chosen to illustrate how to move large amounts of data around quickly and efficiently. The outline of such a test application is shown in the figure below. As can be seen, there are several key aspects involved, but mainly it is about moving large amounts of data around a system quickly, efficiently and reliably.

The basic system is shown in outline form in Figure 9.

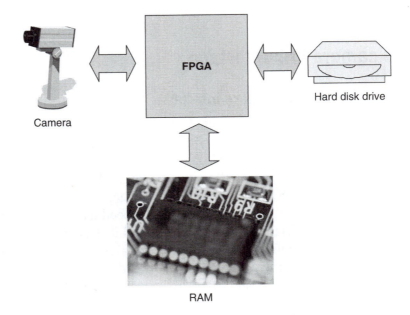

Camera

FPGA

Hard disk drive

RAM

Figure 9
Video Monitor
System Overview

The key performance aspect of this system is in the three interfaces:

1. Camera $\Leftrightarrow$ FPGA
2. 1FPGA $\Leftrightarrow$ PC/Hard disk drive (HDD)
3. FPGA $\Leftrightarrow$ RAM

If we consider the basic camera performance criteria, we have four issues to consider:

1. Resolution
2. Frame rate
3. Color specification
4. Clip size

In this example, the resolution is defined as being 640×480 pixels, the color mode is 24-bit color (3×8 bit planes), the maximum frame rate is 100 s and finally the basic clip size is anything up to 10 s.

What is not shown in the overview figure above is the requirement for some basic control options (such as 'play', 'record', 'store') to allow the stored clips to be replayed using a standard Video Graphics Array (VGA) output (available on most FPGA development kits) or stored for long-term storage on an HDD (or similar high-capacity storage device). This could be handled separately using a PC interface, but that detail is beyond the scope of this basic system description.

The camera link interface

Hardware interface

There are a number of approaches for linking cameras for the high-speed transfer of data, with the two most common being Universal Serial Bus USB (to PCs) and a standard Camera Link using Low Voltage Differential Swing (LVDS) serial data transmission. The LVDS system is a differential serial link that uses voltages of about 350 mV to transmit high-speed data with low noise and low power. Many FPGA development kits have a standard LVDS bus available and this means that the signals can be connected directly between the camera and the FPGA board to transfer data from the camera to the FPGA and hence to the storage (either RAM or HDD).

Data rates

The actual data rate required is theoretically the resolution multiplied by the frame rate multiplied by the number of bits required for each pixel, which in this example would mean the following calculation:

$$\text{Data rate} = \text{Resolution} * \text{frame rate} * \text{bits/pixel} \qquad (1)$$

Which for the specification would mean a total data rate of:

$$\text{Data rate} = 640 * 480 * 100 * 24 \qquad (2)$$

$$\text{Data rate} = 737\,280\,000\,\text{bps} \qquad (3)$$

This equates to a data rate of over 90 MB/s (M bytes per second) and as such is extremely fast for a practical application. Even if the FPGA could run at 100 MHz, the margin on such a system is pretty small.

The Bayer pattern

Luckily, in practice, most camera systems do not use 24 bits in this raw fashion. Kodak have developed the Bayer pattern which is a technique whereby instead of requiring each pixel to have its own individual 3 color planes (requiring 24 bits in total), an array of color filters is placed over the camera sensor and this limits the requirement for data bits per pixel to a single 8-bit byte (with a known color filter in place). The Bayer pattern is repeated over the image in a fixed configuration to standardize this process. The Bayer pattern is shown in Figure 10.

Clearly, using this approach, the required data rate can be divided by three and reduces to a more manageable 30 MB/s.

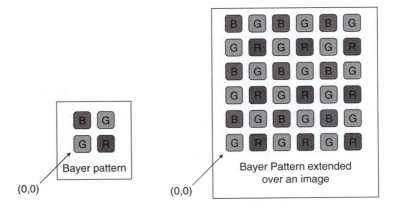

Figure 10
Basic Bayer Pattern, and Extended Over a Larger Image Area

However, the disadvantage of this approach is that the resolution is reduced, however, most images can be reconstructed fairly readily using a method of interpolation which checks firstly which color the current pixel is (red, green or blue) and then takes an average of the neighboring pixels of the missing colors. For example, if the current pixel color is green, then the blue and red color of the current pixel is obtained by averaging the neighboring blue (2) and red (2) pixels, respectively.

Memory requirements

Taking the use of Bayer patterns to reduce the sheer amount of data required into account, this means that the RAM requirements are still high, in this case for a 640×480 image size, this will require a memory size of:

$$\text{Memory size} = \text{resolution} * \text{bits/pixel}$$

$$\text{Memory size} = \text{resolution} * 8 \text{ bits}$$

$$\text{Memory size} = 640 * 480 * 8 \text{ bits}$$

$$\text{Memory size} = 307200 * 8 \text{ bits (per frame)}$$

Clearly, a large memory is going to be required for any significant memory storage and it is unlikely to be possible to store this on the FPGA itself. A more practical solution will be to use some RAM connected to the FPGA (or perhaps available on the development board itself). Options for the memory could include Synchronous Dynamic Random Access Memory (SDRAM) or Flash Memory. Both of these options will be discussed in detail later in the book, however it is useful to consider the advantages and disadvantages of each approach in general.

If we consider SDRAM, the key aspects of this type of memory to consider are:

1. This type of DRAM (Dynamic RAM) relies on transistor capacitance on gates to store data.

2. DRAM is much more compact than SRAM (Static RAM).

3. DRAM cannot be synthesized – you need a separate DRAM chip.

4. SDRAM requires a synchronization clock that is consistent with the rest of the hardware system (it is designed to operate with microprocessors).

5. DRAM data must be refreshed as it is stored charge and decays after a certain time.

6. DRAM is slower than SRAM.

Static RAM (SRAM) can be considered in a similar way to a Read Only Memory (ROM) chip and it also has (differing) key aspects of behavior to consider:

1. Memory cells are based on standard latches.

2. SRAM is fast.

3. SRAM is less compact than DRAM (or SDRAM).

4. SRAM can be synthesized on an FPGA – so is ideal for small, fast registers or memory blocks.

Static RAM is essentially asynchronous, but can be modified to behave synchronously (as SDRAM is the synchronous equivalent of DRAM), and this is often called Synchronous RAM.

Flash Memory is useful to consider at this point, even though its operation is fundamentally different from the memory types considered thus far, simply because it is easy to use and is commonly available on many FPGA development boards.

Flash Memory is essentially a form of EEPROM (Electrically Programmable ROM) that can be used as a form of persistent RAM. Why persistent? In Flash Memory, the device memory is retained even when the power is removed, so it is often used as a form of ROM, which makes it an interesting memory to use on FPGA systems as it could be used to store the FPGA program, but also used as a RAM storage (dynamically) for current data.

Getting started

Now that the basic context of the design has been described, and the basic specification firmed up, the first stage of the actual design can start. In practice, many of the individual blocks may exist in some form, but may need to be modified to fit the specific application requirements. However, generally speaking it is sensible to start with a top-down design methodology.

What that means is that based on the specification, a top level block can be designed that has the correct pin interface (although this may change as the design is refined) and an outline block structure that contains the functional blocks in the design. If we

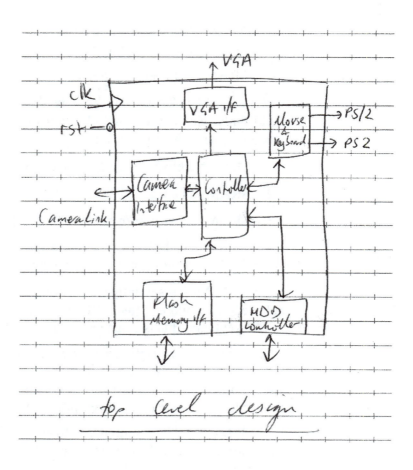

Figure 11
Top Level Design –
Sketch

consider the design example in this section of the book a typical starting point will be a top level diagram showing the basic building blocks of the design and the overall interfaces. Some of the details will not be complete at this stage, but we can start to construct a top level design that we can 'fill in' the details later as we go on with the details of each design block.

Figure 11 shows the outline top level design for the application.

The essential features of the design are captured in this sketch – the main functional blocks, the key interfaces and also notice that we have identified a system clock and reset that will propagate to all the individual functional blocks.

Notice also, that in the original design we did not specify the user input mechanism (i.e. how does the user control the camera interface or store data?). We have taken a design decision at this point, which is to use a simple mouse and keyboard interface to provide the user control to the FPGA system. This allows a flexible approach, so in the first instance, we could use mouse keys or specific keys on the keyboard to initiate a record sequence, or playback, or store, but ultimately, depending on how complex we wish to make the design, it would be possible to design a simple user interface with buttons or similar user interface features, actually on the display to allow controls to drive the system.

Specifying the interfaces

From the sketch shown in Figure 11 we can begin to identify the interface requirements for the top level design. First we clearly need a clock and reset (active low), so keeping things simple (always a good strategy) we can define the clock pin as **clk** and the reset pin as **nrst**. These are standard logic connections, and so we will use the basic standard logic type defined in the IEEE std_logic library. This does not define any details about the actual implementation of the pins (5 V or 3.3 V or even 1 V), but simply the number of logic levels in the model. The actual implementation is defined by the FPGA being used.

Defining the top level design

For this design we must define a top level entity name, and also individual block names. It is always a good idea to use meaningful names (unless they become unmanageable in which case acronyms can be helpful), and hierarchy can also help in keeping duplicate name problems to a minimum. For example, in this case, the design is designed as an image handler and storage interface, which is clearly a handful, so in this example, we will shorten to IHSI (remember that VHDL is case **in**sensitive). Each main block below this top level will then have the prefix ihsi_ to identify the block in the design. This also has the effect of keeping all the blocks grouped together in the same place alphabetically in the compiled library, which makes things easier to find. We can therefore produce the first initial top level entity for the complete application:

```
library ieee;
use ieee.std_logic_1164.all;
entity ihsi is
     Port (
             clk : IN std_logic;
             nrst : IN std_logic
     );
end entity ihsi;
```

We can then identify each major block that requires an external interface and add the requisite connection points to the top level entity. It is worth remembering that at each stage of the design, we do not need to have every block defined completely to test other parts of the design. We can use behavioral models or even empty models to simply ensure that the interfaces are in place and then replace each empty block with a fully functional one. We can also start with behavioral models, replace with Register Transfer Level (RTL) models and finally even replace these with synthesized ones. Thus a complete system can be tested piece by piece until all the blocks are in place.

System block definitions and interfaces

Overall system decomposition

In this specific application we have several important blocks with external interfaces including:

- Mouse controller (PS/2)
- Keyboard controller (PS/2)
- Flash memory
- VGA output
- CameraLink
- PC interface

We can take each of these interfaces in turn and specify the requisite interface connections required for the design.

Mouse and keyboard interfaces

The mouse and keyboard PS/2 interfaces are relatively easy. Each of these has a clock and a data connection and so for each we can define two pins as follows:

Mouse: **mouse_clk, mouse_data**
Keyboard: **key_clk, key_data**

In the general case, the PS/2 interface (to be covered in more detail in the design toolbox section of this book in the PS/2 Mouse and PS/2 Keyboard Chapters [150–160]) allows both directions to be used (i.e. device to controller and vice versa) so these connections must be defined as INOUT std_logic connections in our top level entity.

Memory interface

For the memory interface, we have two options. The first option is to define precisely the type of memory we are going to use in this application (RAM, Flash, EEROM, DRAM, SRAM) and produce a specific interface that will work for only that type of memory. Another approach is to consider that we will treat whatever type of memory we have as generic RAM internally, and to design a memory block that will interface to the actual memory (i.e. we will treat the memory interface as essentially a virtual RAM block). For the initial design, therefore, we can treat the memory as a simple synchronous RAM block that has a clock, data bus, address bus, and write and read signals. For this initial interface, therefore, we will require the following signals only:

Signal	Name	Direction	Type	Notes
Clock	**mem_clk**	OUT	std_logic	
Data bus	**mem_data(31:0)**	INOUT	std_logic	
Address bus	**mem_addr(31:0)**	OUT	std_logic	
Write	**mem_nwr**	OUT	std_logic	(active low)
Read	**mem_nrd**	OUT	std_logic	(active low)

More details on modeling the memory interface and dedicated memory itself is given in the chapter on memory later in this book [140–149].

The display interface: VGA

For the VGA output (to be described later in this book in more detail) we require a specific definition of pins for the connection to the VGA connector on a development board or system. The first set of pins required in any VGA system is the clock and sync pins.

The global VGA clock needs to be set to a specific frequency (depending on the monitor), such as 25 MHz, and this must be derived from the system clock on the FPGA board (say 100 MHz). The VGA clock pin is called the pixel clock and we can use the naming convention of vga_ as a prefix, followed by the functional name. So, for the pixel clock, the pin is named vga_out_pixel_clock.

In addition to the clock, there are three synchronization signals required, the horizontal sync (vga_hsync), the vertical sync (vga_vsync) and the composite sync (vga_comp_sync). Finally, there is a blank pulse (vga_out_blank_z).

The set of pins defined next are the three color data sets. VGA has three color planes (red, green and blue) each with a definition of 8 bits, giving 24 bits in total. As has been described previously, these can be processed using a Bayer pattern, but when the final output pixel data is put together, all three planes require some output values to be set (even if they are all zero). We can define these pins as 8-bit vectors as follows:

```
vga_out_red : OUT std_logic_vector (7 downto 0);
vga_out_green : OUT std_logic_vector (7 downto 0);
vga_out_blue : OUT std_logic_vector (7 downto 0);
```

This provides a complete definition of the VGA interface to the monitor from the system as a whole. More details of the VGA interface mechanism is given in the dedicated chapter on VGA in the designer's toolbox section later in this book [161–168].

The cameralink interface

The cameralink standard has been devised to provide a generic 26 pin interface to a wide range of digital camera and as such we can specify a standard interface at the top level of our design. Although the interface requires 26 pins, they are configured differentially, and so we can specify the basic interface functionally using only 11 pins.

There is a clock pin, which we can define as camera_clk, and then 4 camera control lines defined as cc1 to cc4, respectively. Using the 'camera_' prefix, we can therefore name these as camera_cc1, camera_cc2, camera_cc3 and camera_cc4. There are two serial communication lines serTFG (comms to frame grabber) and serTC (comms to camera) which we can name as camera_sertfg and camera_sertc, respectively. Finally, we have the 4 connection pins from the camera which will contain the data from the device and these are named camera_x0, camera_x1, camera_x2 and camera_x3.

Clearly, the actual interface requires differential outputs, and so eventually an extra interface will be required to translate the simple form of interface defined here to the specific pins of the connector.

The PC interface

The interface to the PC could be using either a standard serial interface such as USB (covered later in this book [93–95]) or using a direct interface to an HDD.

The HDD interface offers a different challenge to the RAM memory interface discussed previously. There are numerous standards for interfacing to HDDs including the major two in current use IDE/AT (Intelligent Drive Electronics/Advanced Technology) and SCSI (Small Computers System Interface). SCSI is commonly used for high-speed drives and has been historically used extensively in Unix-based systems. SCSI is a generic systems interface, and therefore it allows almost ANY type of device to be attached to the system (SCSI) bus. The IDE/AT standard was devised for HDDs only and so has the advantage of being specifically designed for HDD interfaces. IDE (Intelligent Drive Electronics/AT Attachment) drives are generally slower, but significantly cheaper than SCSI drives and so PCs tend to use an IDE/ATA interface and higher end workstations will use SCSI drives instead.

In this context, the IDE/ATA drive is highly appropriate as the interface is much simpler than the SCSI interface, and therefore more practical in developing a prototype system. If a more advanced system is required, then clearly this can be changed later. The IDE approach is to have a number of master and slave devices on the bus (anyone who has looked inside a PC will recognize the need for setting a master/slave switch or jumper on a drive before installation of an extra or new HDD). A bus controller sets a series of registers with commands and the selected device on the chain will execute. It is worth noting that the bus will operate at the speed of the slowest device on the chain.

There are a total of 13 registers in the IDE/ATA configuration. These registers are divided into command block registers and control block registers. The command block registers are for sending commands to the device or for posting the status of the device. The control block registers are used for device control and for posting an alternate status. The full details of interfacing to an IDE/ATA device is beyond the scope of this book and is not used in this example.

The complexity of the IDE/ATA interface is such that it would probably take several thousand lines of VHDL to implement completely. If the performance requirements were such that it was essential, then the reader can find numerous sources of information to implement this design, including the ATA 6/UDMA100 specification.

An alternative approach is to use a standard interface such as USB with memory buffering and compression to manage the data storage issues, where the USB interface is discussed in detail in the designer's toolbox part of this book [93–96].

Summary

In summary, this chapter shows how a high-level specification can be practically decomposed into a series of manageable problems that may all have a relatively simple solution. The key to successful systems design is to decompose the design into blocks that have a definable core function. This can then be implemented directly in VHDL. The second aspect of the design is to analyze the boundaries.

A common phrase coined by systems designers is 'problems migrate to the boundaries'. In other words, we can easily construct a VHDL design if we know the core functionality, however, getting the individual blocks to communicate successfully is often much harder. As a result, the designer often spends a lot of debug time in integrating a number of different functions together, and being forced to rewrite large sections of code to make that happen.

A useful approach to handling this specific problem is to create 'empty' VHDL models that do not operate functionally, but *do* have the correct interfaces. These models can be tested with basic communications test data to ensure that the correct signals are in place, the data can be passed around the complete design at the required data rates, and that errors in signal names, directions and types can be sorted out prior to developing the core VHDL.

Hopefully this chapter has provided a useful introduction to modeling and designing complex systems using VHDL and the general approach of thinking at a high level without going too deeply into the details of each block has been highlighted as a useful approach.

6

Embedded Processors

Introduction

This application example chapter concentrates on the key topic of Integrating Processors onto Field Programmable Gate Array (FPGA) designs. This ranges from simple 8-bit microprocessors up to large IP processor cores that require an element of hardware–software co-design involved. This chapter will take the reader through the basics of implementing a behavioral-based microprocessor for evaluation of algorithms, through to the practicalities of structurally correct models that can be synthesised and implemented on an FPGA.

One of the major challenges facing hardware designers in the 21st century is the problem of hardware–software co-design. This has moved on from a basic partitioning mechanism based on standard hardware architectures to the current situation where the algorithm itself can be optimized at a compilation level for performance or power by implementing appropriately at different levels with hardware or software as required.

This aspect suits FPGAs perfectly, as they can handle fixed hardware architecture that runs software compiled onto memory, they can implement optimal hardware running at much faster rates than a software equivalent could, and there is now the option of configurable hardware that can adapt to the changing requirements of a modified environment.

A simple embedded processor

Embedded processor architecture

A useful example of an embedded processor is to consider a generic microcontroller in the context of an FPGA platform. Take

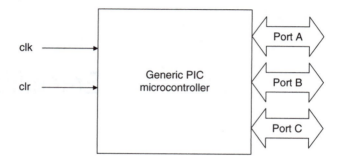

Figure 12
Simple
Microcontroller

a simple example of a generic 8-bit microcontroller shown in Figure 12.

As can be seen from Figure 12, the microcontroller is a 'general purpose microprocessor', with a simple clock (clk) and reset (clr), and three 8-bit ports (A, B and C). Within the microcontroller itself, there needs to be the following basic elements:

1. A control unit: This is required to manage the clock and reset of the processor, manage the data flow and instruction set flow, and control the port interfaces. There will also need to be a Program Counter (PC).

2. An Arithmetic Logic Unit (ALU): a PIC will need to be able to carry out at least some rudimentary processing – carried out in the ALU.

3. An address bus.

4. A data bus.

5. Internal registers.

6. An instruction decoder.

7. A Read Only Memory (ROM) to hold the program.

While each of these individual elements (1–6) can be implemented simply enough using a standard FPGA, the ROM presents a specific difficulty. If we implement a ROM as a set of registers, then obviously this will be hugely inefficient in an FPGA architecture. However, in most modern FPGA platforms, there are blocks of Random Access Memory (RAM) on the FPGA that can be accessed and it makes a lot of sense to design a RAM block for use as a ROM by initializing it with the ROM values on reset and then using that to run the program.

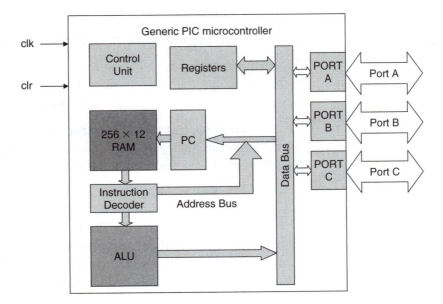

Figure 13
Embedded
Microcontroller
Architecture

This aspect of the embedded core raises an important issue, which is the reduction in efficiency of using embedded rather than dedicated cores. There is usually a compromise involved and in this case it is that the ROM needs to be implemented in a different manner, in this case with a hardware penalty. The second issue is what type of memory core to use? In an FPGA RAM, the memory can usually be organized in a variety of configurations to vary the depth (number of memory addresses required) and the width (width of the data bus). For example a 512 address RAM block, with an 8-bit address width would be equivalent to a 256 address RAM block with a 16-bit address width.

If the equivalent ROM is, say 12 bits wide and 256, then we can use the 256×16 RAM block and ignore the top four bits. The resulting embedded processor architecture could be of the form shown in Figure 13.

Basic instructions

When we program a microprocessor of any type, there are three different ways of representing the code that will run on the processor. These are machine code (1's and 0's), assembler (low-level instructions such as LOAD, STORE, . . .) and high-level code (such as C, Fortran or Pascal). Regardless of the language used, the code will always be compiled or assembled into machine code at the lowest level for programming into memory. High-level code (e.g. C) is compiled and assembler code is assembled (as the name suggests) into machine code for the specific platform.

Clearly a detailed explanation of a compiler is beyond the scope of this book, but the same basic process can be seen in an assembler and this is useful to discuss in this context.

Every processor has a basic 'Instruction Set' which is simply the list of functions that can be run in a program on the processor. Take the simple example of the following pseudocode expression:

$$b = a + 2$$

In this example, we are taking the variable a and adding the integer value 2 to it, and then storing the result in the variable b. In a processor, the use of a variable is simply a memory location that stores the value, and so to load a variable we use an assembler command as follows:

LOAD a

What is actually going on here? Whenever we retrieve a variable value from memory, the implication is that we are going to put the value of the variable in the register called the accumulator (ACC). The command 'LOAD a' could be expressed in natural language as 'LOAD the value of the memory location denoted by a into the accumulator register ACC'.

The next stage of the process is to add the integer value 2 to the accumulator. This is a simple matter, as instead of an address, the value is simply added to the current value stored in the accumulator. The assembly language command would be something like:

ADD #x02

Notice that we have used the x to denote a hexadecimal number. If we wished to add a variable, say called c, then the command would be the same, except that it would use the address c instead of the absolute number. The command would therefore be:

ADD c

Now we have the value of a + 2 stored in the accumulator register (ACC). This could be stored in a memory location, or put onto a port (e.g. PORT A). It is useful to notice that for a number we use the key character # to indicate that we are adding the value and not using the argument as the address.

In the pseudocode example, we are storing the result of the addition in the variable called b, so the command would be something like this:

STORE b

While this is superficially a complete definition of the instruction set requirements, there is one specific design detail that has to be decided on for any processor. This is the number of instructions and the data bus size. If we have a set of instructions with the number of instructions denoted by I, then the number of bits in the opcode (n) must conform to the following rule:

$$2^n \leq I \tag{4}$$

In other words, the number of bits provides the number of unique different codes that can be defined, and this defines the size of the instruction set possible. For example, if $n = 3$, then with three bits there are eight possible unique opcodes, and so the maximum size of the instruction set is eight.

Fetch execute cycle

The standard method of executing a program in a processor is to store the program in memory and then follow a strict sequence of events to carry out the instructions. The first stage is to use the PC to increment the program line, this then calls up the next command from memory in the correct order, and then the instruction can be loaded into the appropriate register for execution. This is called the 'fetch execute cycle'.

What is happening at this point? First the contents of the PC is loaded into the Memory Address Register (MAR). The data in the memory location are then retrieved and loaded into the Memory Data Register (MDR). The contents of the MDR can then be transferred into the Instruction Register (IR). In a basic processor, the PC can then be incremented by one (or in fact this could take place immediately after the PC has been loaded into the MDR).

Once the opcode (and arguments if appropriate) are loaded, then the instruction can be executed. Essentially, each instruction has its own state machine and control path, which is linked to the IR and a sequencer that defines all the control signals required to move the data correctly around the memory and registers for that instruction. We will discuss registers in the next section, but in addition to the

PC, IR and accumulator (ACC) mentioned already, we require two memory registers as a minimum, the MDR and MAR.

For example, consider the simple command LOAD a, from the previous example. What is required to actually execute this instruction? First, the opcode is decoded and this defines that the command is a 'LOAD' command. The next stage is to identify the address. As the command has not used the # symbol to denote an absolute address, this is stored in the variable 'a'. The next stage, therefore is to load the value in location 'a' into the MDR, by setting MAR = a and then retrieving the value of a from the RAM. This value is then transferred to the accumulator (ACC).

Embedded processor register allocation

The design of the registers partly depends on whether we wish to 'clone' a PIC device or create a modified version that has more custom behavior. In either case there are some mandatory registers that must be defined as part of the design. We can assume that we need an accumulator (ACC), a Program Counter (PC), and the three input/output ports (PORTA, PORTB, PORTC). Also, we can define the IR, MAR, and MDR.

In addition to the data for the ports, we need to have a definition of the port direction and this requires three more registers for managing the tristate buffers into the data bus to and from the ports (DIRA, DIRB, DIRC). In addition to this, we can define a number (essentially arbitrary) of registers for general purpose usage. In the general case the naming, order and numbering of registers does not matter, however, if we intend to use a specific device as a template, and perhaps use the same bit code, then it is vital that the registers are configured in exactly the same way as the original device and in the same order.

In this example, we do not have a base device to worry about, and so we can define the general purpose registers (24 in all) with the names REG0 to REG23. In conjunction with the general purpose registers, we need to have a small decoder to select the correct register and put the contents onto the data bus (F).

A basic instruction set

In order for the device to operate as a processor, we must define some basic instructions in the form of an instruction set. For this simple example we can define some very basic instructions that will carry out basic program elements, ALU functions, memory functions. These are summarised in the following table:

Command	Description

LOAD arg	This command loads an argument into the accumulator. If the argument has the prefix # then it is the absolute number, otherwise it is the address and this is taken from the relevant memory address.
	
	Examples:
	
	   LOAD #01
	
	   LOAD abc

STORE arg	This command stores an argument from the accumulator into memory. If the argument has the prefix # then it is the absolute address, otherwise it is the address and this is taken from the relevant memory address.
	
	Examples:
	
	   STORE #01
	
	   STORE abc

ADD arg	This command adds an argument to the accumulator. If the argument has the prefix # then it is the absolute number, otherwise it is the address and this is taken from the relevant memory address.
	
	Examples:
	
	   ADD #01
	
	   ADD abc

NOT	This command carries out the NOT function on the accumulator.

AND arg	This command ands an argument with the accumulator. If the argument has the prefix # then it is the absolute number, otherwise it is the address and this is taken from the relevant memory address.
	
	Examples:
	
	   AND #01
	
	   AND abc

OR arg	This command ors an argument with the accumulator. If the argument has the prefix # then it is the absolute number, otherwise it is the address and this is taken from the relevant memory address.
	
	Examples:
	
	   OR #01
	
	   OR abc

(continued)

Table (*continued*)	
Command	**Description**
XOR arg	This command xors an argument with the accumulator. If the argument has the prefix # then it is the absolute number, otherwise it is the address and this is taken from the relevant memory address. Examples: XOR #01 XOR abc
INC	This command carries out an increment by one on the accumulator.
SUB arg	This command subtracts an argument from the accumulator. If the argument has the prefix # then it is the absolute number, otherwise it is the address and this is taken from the relevant memory address. Examples: SUB #01 SUB abc
BRANCH arg	This command allows the program to branch to a specific point in the program. This may be very useful for looping and program flow. If the argument has the prefix # then it is the absolute number, otherwise it is the address and this is taken from the relevant memory address. Examples: BRANCH #01 BRANCH abc

In this simple instruction set, there are 10 separate instructions. This implies, from the rule given in equation (4) previously in this chapter, that we need at least 4 bits to describe each of the instructions given in the table above. Given that we wish to have 8 bits for each data word, we need to have the ability to store the program memory in a ROM that has words of at least 12 bits wide. In order to cater for a greater number of instructions, and also to handle the situation for specification of different addressing modes (such as the difference between absolute numbers and variables), we can therefore suggest a 16-bit system for the program memory.

Notice that at this stage there are no definitions for port interfaces or registers. We can extend the model to handle this behavior later.

Structural or behavioral?

So far in the design of the simple microprocessor, we have not specified detailed beyond a fairly abstract structural description of the processor in terms of registers and busses. At this stage we have a decision about the implementation of the design with regard to the program and architecture.

One option is to take a program (written in assembly language) and simply convert this into a state machine that can easily be implemented in a VHDL model for testing out the algorithm. Using this approach, the program can be very simply modified and recompiled based on simple rules that restrict the code to the use of registers and techniques applicable to the processor in question. This can be useful for investigating and developing algorithms, but is more ideal than the final implementation as there will be control signals and delays due to memory access in a processor plus memory configuration, that will be better in a dedicated hardware design.

Another option is to develop a simple model of the processor that does have some of the features of the final implementation of the processor, but still uses an assembly language description of the model to test. This has advantages in that no compilation to machine code is required, but there are still not the detailed hardware characteristics of the final processor architecture that may cause practical issues on final implementation.

The third option is to develop the model of the processor structurally and then the machine code can be read in directly from the ROM. This is an excellent approach that is very useful for checking both the program and the possible quirks of the hardware/software combination as the architecture of the model reflects directly the structure of the model to be implemented on the FPGA.

Machine code instruction set

In order to create a suitable instruction set for decoding instructions for our processor, the assembly language instruction set needs to have an equivalent machine code instruction set that can be decoded by the sequencer in the processor. The resulting opcode/instruction table is given below:

Command	Opcode (Binary)
LOAD arg	0000
STORE arg	0001
ADD arg	0010
NOT	0011
AND arg	0100
OR arg	0101
XOR arg	0110
INC	0111
SUB arg	1000
BRANCH arg	1001

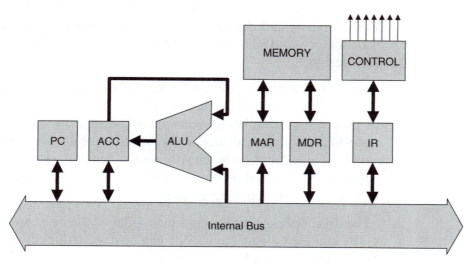

Figure 14
Structural Model of the Microprocessor

Structural elements of the microprocessor

Taking the abstract design of the microprocessor given in Figure 13 we can redraw with the exact registers and bus configuration as shown in the structural diagram in Figure 14. Using this model we can create separate VHDL models for each of the blocks that are connected to the internal bus and then design the control block to handle all the relevant sequencing and control flags to each of the blocks in turn.

Before this can be started, however, it makes sense to define the basic criteria of the models and the first is to define the basic type. In any digital model (as we have seen elsewhere in this book) it is sensible to ensure that data can be passed between standard models and so in this case we shall use the std_logic_1164 library that is the standard for digital models.

In order to use this library, each signal shall be defined as of the basic type std_logic and also the library ieee.std_logic_1164.all shall be declared in the header of each of the models in the processor.

Finally, each block in the processor shall be defined as a separate block for implementation in VHDL.

Processor functions package

In order to simplify the VHDL for each of the individual blocks, a set of standard functions have been defined in a package call processor_functions. This is used to define useful types and functions for this set of models. The VHDL for the package is given below:

```
Library ieee;
Use ieee.std_logic_1164.all;

Package processor_functions is
      Type opcode is (load, store, add, not, and, or,
        xor, inc, sub, branch);
      Function Decode (word : std_logic_vector) return
        opcode;
      Constant n : integer := 16;
      Constant oplen : integer := 4;
      Type memory_array is array (0 to 2**(n-oplen-1)
        of Std_logic_vector(n-1 downto 0);
      Constant reg_zero : unsigned (n-1 downto 0)  :=
        (others => '0');
End package processor_functions;

Package body processor_functions is
      Function Decode (word : std_logic_vector) return
        opcode is
            Variable opcode_out : opcode;
```

```
      Begin
          Case word(n-1 downto n-oplen-1) is
                When "0000" => opcode_out : = load;
                When "0001" => opcode_out : = store;
                When "0010" => opcode_out : = add;
                When "0011" => opcode_out : = not;
                When "0100" => opcode_out : = and;
                When "0101" => opcode_out : = or;
                When "0110" => opcode_out : = xor;
                When "0111" => opcode_out : = inc;
                When "1000" => opcode_out : = sub;
                When "1001" => opcode_out : = branch;
                When others => null;
          End case;
          Return opcode_out;
      End function decode;
End package body processor_functions;
```

The PC

The PC needs to have the system clock and reset connections, the system bus (defined as inout so as to be readable and writable by the PC register block). In addition, there are several control signals required for correct operation. The first is the signal to increment the PC (PC_inc), the second is the control signal load the PC with a specified value (PC_load) and the final is the signal to make the register contents visible on the internal bus (PC_valid). This signal ensures that the value of the PC register will appear to be high impedance ('Z') when the register is not required on the processor bus. The system bus (PC_bus) is defined as a std_logic_vector, with direction inout to ensure the ability to read and write. The resulting VHDL entity is given below:

```
library ieee;
use ieee.std_logic_1164.all;
entity pc is
    Port (
        Clk : IN std_logic;
        Nrst : IN std_logic;
        PC_inc : IN std_logic;
        PC_load : IN std_logic;
        PC_valid : IN std_logic;
        PC_bus : INOUT std_logic_vector(n-1 downto 0)
    );
End entity PC;
```

The architecture for the PC must handle all of the various configurations of the PC control signals and also the communication of the data into and from the internal bus correctly. The PC model has an asynchronous part and a synchronous section. If the PC_valid goes low at any time, the value of the PC_bus signal should be set to 'Z' across all of its bits. Also, if the reset signal goes low, then the PC should reset to zero.

The synchronous part of the model is the increment and load functionality. When the clk rising edge occurs, then the two signals PC_load and PC_inc are used to define the function of the counter. The precedence is that if the increment function is high, then regardless of the load function, then the counter will increment. If the increment function (PC_inc) is low, then the PC will load the current value on the bus, if and only if the PC_load signal is also high.

The resulting VHDL is given below:

```
architecture RTL of PC is
      signal counter : unsigned (n-1 downto 0);
begin
      PC_bus <= std_logic_vector(counter)
                  when PC_valid = '1' else (others =>
                  'Z');
      process (clk, nrst) is
      begin
            if nrst = '0' then
                      count <= 0;
            elsif rising_edge(clk) then
                      if PC_inc = '1' then
                          count <= count + 1;
                      else
                              if PC_load = '1' then
                                  count <= unsigned(PC_bus);
                              end if;
                      end if;
            end if;
      end process;
end architecture RTL;
```

The IR

The IR has the same clock and reset signals as the PC, and also the same interface to the bus (IR_bus) defined as a std_logic_vector of type INOUT. The IR also has two further control signals, the first being the command to load the IR (IR_load), and the second being to load the required address onto the system bus (IR_address). The final connection is the decoded opcode that is to be sent to the system controller. This is defined as a simple unsigned integer value with the same size as the basic system bus. The basic VHDL for the entity of the IR is given below:

```
library ieee;
use ieee.std_logic_1164.all;
use work.processor_functions.all;
entity ir is
      Port (
            Clk : IN std_logic;
            Nrst : IN std_logic;
            IR_load : IN std_logic;
            IR_valid : IN std_logic;
            IR_address : IN std_logic;
            IR_opcode : OUT opcode;
            IR_bus : INOUT std_logic_vector(n-1 downto 0)
      );
End entity IR;
```

The function of the IR is to decode the opcode in binary form and then pass to the control block. If the IR_valid is low, the bus value should be set to 'Z' for all bits. If the reset signal (nsrt) is low, then the register value internally should be set to all 0's.

On the rising edge of the clock, the value on the bus shall be sent to the internal register and the output opcode shall be decoded asynchronously when the value in the IR changes.

The resulting VHDL architecture is given below:

```
architecture RTL of IR is
      signal IR_internal : std_logic_vector (n-1 downto 0);
begin
      IR_bus <= IR_internal
            when IR_valid = '1' else (others => 'Z');
      IR_opcode <= Decode(IR_internal);
      process (clk, nrst) is
      begin
            if nrst = '0' then
                  IR_internal <= (others => '0');
            elsif rising_edge(clk) then
                  if IR_load = '1' then
                        IR_internal <= IR_bus;
                  end if;
            end if;
      end process;
end architecture RTL;
```

In this VHDL, notice that we have used the predefined function Decode from the processor_functions package previously defined. This will look at the top four bits of the address given to the IR and decode the relevant opcode for passing to the controller.

The Arithmetic and Logic Unit

The Arithmetic and Logic Unit (ALU) has the same clock and reset signals as the PC, and also the same interface to the bus (ALU_bus) defined as a std_logic_vector of type INOUT. The ALU also has three further control signals, which can be decoded to map to the eight individual functions required of the ALU. The ALU also contains the Accumulator (ACC) which is a std_logic_vector of the size defined for the system bus width. There is also a single-bit output ALU_zero which goes high when all the bits in the accumulator are zero.

The basic VHDL for the entity of the ALU is given below:

```
library ieee;
use ieee.std_logic_1164.all;
use work.processor_functions.all;
entity alu is
      Port (
            Clk : IN std_logic;
            Nrst : IN std_logic;
            ALU_cmd : IN std_logic_vector(2 downto 0);
            ALU_zero : OUT std_logic;
            ALU_valid : IN std_logic;
            ALU_bus : INOUT std_logic_vector(n-1 downto 0)
      );
End entity alu;
```

The function of the ALU is to decode the ALU_cmd in binary form and then carry out the relevant function on the data on the bus, and the current data in the accumulator. If the ALU_valid is low, the bus value should be set to 'Z' for all bits. If the reset signal (nsrt) is low, then the register value internally should be set to all 0's.

On the rising edge of the clock, the value on the bus shall be sent to the internal register and the command shall be decoded.

The resulting VHDL architecture is given below:

```
architecture RTL of ALU is
      signal ACC : std_logic_vector (n-1 downto 0);
begin
      ALU_bus <= ACC
            when ACC_valid = '1' else (others => 'Z');
      ALU_zero <= '1' when acc = reg_zero else '0';
      process (clk, nrst) is
      begin
            if nrst = '0' then
                  ACC <= (others => '0');
```

```
            elsif rising_edge(clk) then
                case ACC_cmd is
                -- Load the Bus value into the
                    accumulator
                when "000" => ACC <= ALU_bus;
                -- Add the ACC to the Bus value
                When "001" => ACC <= add(ACC,ALU_bus);
                -- NOT the Bus value
                When "010" => ACC <= NOT ALU_bus;
                -- OR the ACC to the Bus value
                When "011" => ACC <= ACC or ALU_bus;
                -- AND the ACC to the Bus value
                When "100" => ACC <= ACC and ALU_bus;
                -- XOR the ACC to the Bus value
                When "101" => ACC <= ACC xor ALU_bus;
                -- Increment ACC
                When "110" => ACC <= ACC + 1;
                -- Store the ACC value
                When "111" => ALU_bus <= ACC;
            end if;
        end process;
end architecture RTL;
```

The memory

The processor requires a RAM memory, with an address register (MAR) and a data register (MDR). There therefore needs to be a load signal for each of these registers: MDR_load and MAR_load. As it is a memory, there also needs to be an enable signal (M_en), and also a signal denote Read or Write modes (M_rw). Finally, the connection to the system bus is a standard inout vector as has been defined for the other registers in the microprocessor.

The basic VHDL for the entity of the memory block is given below:

```
library ieee;
use ieee.std_logic_1164.all;
use work.processor_functions.all;
entity memory is
    Port (
        Clk : IN std_logic;
        Nrst : IN std_logic;
        MDR_load : IN std_logic;
        MAR_load : IN std_logic;
        MAR_valid : IN std_logic;
        M_en : IN std_logic;
        M_rw : IN std_logic;
        MEM_bus : INOUT std_logic_vector(n-1
          downto 0)
    );
End entity memory;
```

The memory block has three aspects. The first is the function that the memory address is loaded into the MAR. The second function is either reading from or writing to the memory using the MDR. The final function, or aspect of the memory is to store the actual program that the processor will run. In the VHDL model, we will achieve this by using a constant array to store the program values.

The resulting basic VHDL architecture is given below:

```vhdl
architecture RTL of memory is
    signal mdr : std_logic_vector(wordlen-1 downto 0);
    signal mar : unsigned(wordlen-oplen-1 downto 0);
                begin
    MEM_bus <= mdr
            when MEM_valid = '1' else (others => 'Z');
    process (clk, nrst) is
            variable contents : memory_array;
            constant program : contents :=
            (
                    0 => "0000000000000011",
                    1 => "0010000000000100",
                    2 => "0001000000000101",
                    3 => "0000000000001100",
                    4 => "0000000000000011",
                    5 => "0000000000000000" ,
                    Others => (others => '0')
            );
    begin
            if nrst = '0' then
                    mdr <= (others => '0');
                    mdr <= (others => '0');
                    contents := program;
            elsif rising_edge(clk) then
                    if MAR_load = '1' then
                            mar <= unsigned(MEM_bus(n-oplen-
                                    1 downto 0));
                    elsif MDR_load = '1' then
                            mdr <= MEM_bus;
                    elsif MEM_en = '1' then
                            if MEM_rw = '0' then
                                    mdr <= contents(to_integer
                                            (mar));
                            else
                                    mem(to_integer(mar))
                                            := mdr;
                            end if;
                    end if;
            end if;
    end process;
end architecture RTL;
```

We can look at some of the VHDL in a bit more detail and explain what is going on at this stage. There are two internal signals to the block, mdr and mar (the data and address, respectively). The first aspect to notice is that we have defined the MAR as an unsigned rather than as a std_logic_vector. We have done this to make indexing direct. The MDR remains as a std_logic_vector. We can use an integer directly, but an unsigned translates easily into a std_logic_vector.

```
signal mdr : std_logic_vector(wordlen-1 downto 0);
signal mar : unsigned(wordlen-oplen-1 downto 0);
```

The second aspect is to look at the actual program itself. We clearly have the possibility of a large array of addresses, but in this case we are defining a simple three line program:

$$c = a + b$$

The binary code is shown below:

```
0 => "0000000000000011",
1 => "0010000000000100",
2 => "0001000000000101",
3 => "0000000000001100",
4 => "0000000000000011",
5 => "0000000000000000" ,
Others => (others => '0')
```

For example, consider the line of the declared value for address 0. The 16 bits are defined as 0000000000000011. If we split this into the opcode and data parts we get the following:

Opcode   0000

Data   000000000011 (3)

In other words this means LOAD the variable from address 3. Similarly, the second line is ADD from 4, finally the third command is STORE in 5. In addresses 3, 4 and 5, the three data variables are stored.

Microcontroller: controller

The operation of the processor is controlled in detail by the sequencer, or controller block. The function of this part of the processor is to take the current PC address, look up the relevant instruction from memory, move the data around as required, setting up all the relevant control signals at the right time, with the right values.

As a result, the controller must have the clock and reset signals (as for the other blocks in the design), a connection to the global bus and finally all the relevant control signals must be output. An example entity of a controller is given below:

```
library ieee;
use ieee.std_logic_1164.all;
use work.processor_functions.all;
entity controller is
     generic (
             n : integer := 16
     );
     Port (
             Clk : IN std_logic;
             Nrst : IN std_logic;
             IR_load : OUT std_logic;
             IR_valid : OUT std_logic;
             IR_address : OUT std_logic;
             PC_inc : OUT std_logic;
             PC_load : OUT std_logic;
             PC_valid : OUT std_logic;
             MDR_load : OUT std_logic;
             MAR_load : OUT std_logic;
             MAR_valid : OUT std_logic;
             M_en : OUT std_logic;
             M_rw : OUT std_logic;
             ALU_cmd : OUT std_logic_vector(2 downto 0);
             CONTROL_bus : INOUT std_logic_vector(n-1
                downto 0)
     );
End entity controller;
```

Using this entity, the control signals for each separate block are then defined, and these can be used to carry out the functionality requested by the program. The architecture for the controller is then defined as a basic state machine to drive the correct signals. The basic state machine for the processor is defined in Figure 15.

We can implement this using a basic VHDL architecture that implements each state using a new state type and a case statement to manage the flow of the state machine. The basic VHDL architecture is shown below and it includes the basic synchronous

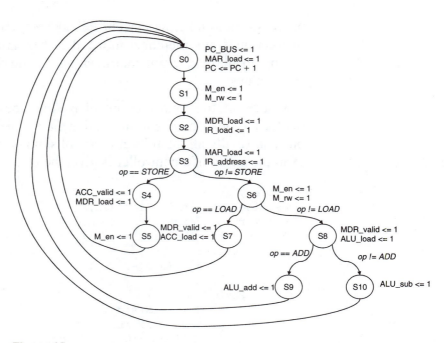

Figure 15
Basic Processor Controller State Machine

machine control section (reset and clock) the management of the next stage logic:

```
architecture RTL of controller is
      type states is
         (s0,s1,s2,s3,s4,s5,s6,s7,s8,s9,s10);
      signal current_state, next_state : states;
begin
         state_sequence: process (clk, nrst) is
               if nrst = '0' then
                     current_state <= s0;
               else
                     if rising_edge(clk) then
                        current_state <=
                           next_state;
                     end if;
               end if;
         end process state_sequence;

         state_machine : process ( present_state,
           opcode ) is
               -- state machine goes here
         End process state_machine;
end architecture;
```

You can see from this VHDL that the first process (state_sequence) manages the transition of the current_state to the next_state and also the reset condition. Notice that this is a synchronous machine and as such waits for the rising_edge of the clock, and that the reset is asynchronous. The second process (state_machine) waits for a change in the state or the opcode and this is used to manage the transition to the next state, although the actual transition itself is managed by the state_sequence process. This process is given in the VHDL below:

```vhdl
state_machine : process ( present_state,
  opcode ) is
begin
    -- Reset all the control signals
    IR_load <= '0';
    IR_valid <= '0';
    IR_address <= '0';
    PC_inc <= '0';
    PC_load <= '0';
    PC_valid <= '0';
    MDR_load <= '0';
    MAR_load <= '0';
    MAR_valid <= '0';
    M_en <= '0';
    M_rw <= '0';
    Case current_state is
    When s0 =>
        PC_valid <= '1'; MAR_load <= '1';
        PC_inc <= '1'; PC_load <= '1';
        Next_state <= s1;
    When s1 =>
        M_en <='1'; M_rw <= '1';
        Next_state <= s2;
    When s2 =>
        MDR_valid <= '1'; IR_load <= '1';
        Next_state <= s3;
    When s3 =>
        MAR_load <= '1'; IR_address <= '1';
        If opcode = STORE then
                Next_state <= s4;
        else
                Next_state <=s6;
        End if;
    When s4 =>
        MDR_load <= '1'; ACC_valid <= '1';
        Next_state <= s5;
    When s5 =>
        M_en <= '1';
        Next_state <= s0;
    When s6 =>
        M_en <= '1'; M_rw <= '1';
```

```
                      If opcode = LOAD then
                             Next_state <= s7;
                      else
                             Next_state <= s8;
                      End if;
               When s7 =>
                      MDR_valid <= '1'; ACC_load <= '1';
                      Next_state <= s0;
               When s8 =>
                      M_en<='1'; M_rw <= '1';
                      If opcode = ADD then
                             Next_state <= s9;
                      else
                             Next_state <= s10;
                      End if;
               When s9 =>
                      ALU_add <= '1';
                      Next_state <= s0;
               When s10 =>
                      ALU_sub <= '1';
                      Next_state <= s0;
          End case;
     End process state_machine;
```

Summary of a simple microprocessor

Now that the important elements of the processor have been defined, it is a simple matter to instantiate them in a basic VHDL netlist and create a microprocessor using these building blocks. It is also a simple matter to modify the functionality of the processor by changing the address/data bus widths or extend the instruction set.

Soft core processors on an FPGA

While the previous example of a simple microprocessor is useful as a design exercise and helpful to gain understanding about how microprocessors operate, in practice most FPGA vendors provide standard processor cores as part of an embedded development kit that includes compilers and other libraries. For example this could be the Microblaze core from Xilinx or the NIOS core supplied by Altera. In all these cases the basic idea is the same, that a standard configurable core can be instantiated in the design and code compiled using a standard compiler and downloaded to the processor core in question.

Each soft core is different and rather than describe the details of a particular case, in this section the general principles will be covered and the reader is encouraged to experiment with the offerings from the FPGA vendors to see which suits their application the best.

In any soft core development system there are several key functions that are required to make the process easy to implement. The first is the system building function. This enables a core to be designed into a hardware system that includes memory modules, control functions, Direct Memory Access (DMA) functions, data interfaces and interrupts. The second is the choice of processor types to implement. A basic NIOS II or similar embedded core will typically have a performance in the region of 100–200 MIPS, and the processor design tools will allow the size of the core to be traded off with the hardware resources available and the performance required.

Summary

The topic of embedded processors on FPGAs would be suitable for a complete book in itself. In this chapter the basic techniques have been described for implementing a simple processor directly on the FPGA and the approach for implementing soft cores on FPGAs have been introduced.

Part 3
Designer's Toolbox

The third section of this book is a designer's toolbox of functions. These are relatively standard functions that occur in many designs and test circuits, and so it is incredibly useful to have at least an initial design to evaluate rather than having to develop one from scratch.

The first chapter in this part (chapter 7) looks at serial communications, starting from the fundamentals of data transmission and discussing a practical approach of incorporating USB into the design. The next chapter (chapter 8) discusses digital filtering, with a simple example to show how to take a standard Laplace (S domain) description and implement it in a VHDL digital filter. Chapter 9 is an introduction to an increasingly important topic – secure systems – with a description of block ciphers, DES and AES. The second half of this section is concerned with standard interfaces. Chapter 10 looks at modeling memory in VHDL, with a description of ROM, RAM, Flash and SRAM. Chapters 11 and 12 describe how to implement a simple PS/2 interface to a mouse and keyboard respectively. The data modes and protocols are reviewed and a simple implementation for each is described. Finally, Chapter 13 shows how to build a simple VGA interface, complete with Sync timing code in VHDL.

This chapter is a useful starting point for those who need to develop complex applications, but often need to build a framework from scratch, but do not wish to develop all the routines from nothing. This part of the book gives a 'helping hand' up the learning curve, although it must be stressed that the examples given are purely for education and as such have been written with simplicity and clarity in mind.

7

Serial Communications

Introduction

There are a wide variety of serial communications protocols available, but all rely on some form of coding scheme to efficiently and effectively transmit the serial data across the transmission medium. In this chapter, not only will the common methods of transmitting data be reviewed (RS-232 and Universal Serial Bus (USB)), but in addition some useful coding mechanisms will be described (Manchester, Code Mark Inversion, Non-Return-toZero – NRZ, Non-Return-toZero-Inverted – NRZI) as they often are used as part of a higher level transmission protocol. For example, the NRZI coding technique is used in the USB protocol.

Manchester encoding and decoding

Manchester encoding is a simple coding scheme that translates a basic bit stream into a series of transitions. It is extremely useful for ensuring that a specific bandwidth can be used for data transmission, as no matter what the sequence of the data bits, the frequency of the transmitted stream will be exactly twice the frequency of the original data. It also makes signal recovery trivial, because there is no need to attempt to extract a clock as the data can be recovered simply by looking for the edges in the data and extracting asynchronously. The basic approach to Manchester encoding is shown in Figure 16.

Another advantage of the scheme is that the method is highly tolerant of errors in the data, if an error occurs, then the subsequent data is not affected at all by an error in the transmitter, the medium or the receiver, and after the immediate glitch, the data can continue to be transmitted effectively without any need

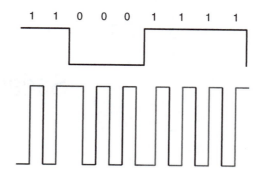

Figure 16
Manchester Encoding
Scheme

for error recovery. Of course, the original data can use some
form of data coding to add in error correction (see the chapter on
data checking on this book for methods such as parity checks,
or CRC).

If we wish to create a VHDL model for this type of coding
scheme, it is actually relatively simple. The first step is to identify
that we have a single data input (D) and a clock (CLK). Why synchronous? Using a synchronous clock we can define a sample on
the rising edge of the clock for the data input and use BOTH edges
of the clock to define the transitions on the output. The resulting
behavioural VHDL code is shown below:

```
Library ieee;
Use ieee.std_logic_1664.all;

Entity Manchester_encoder is
     Port (
             Clk : in std_logic;
             D : in std_logic;
             Q : out std_logic
        );
End entity Manchester_encoder;

Architecture basic of Manchester_encoder is
     Signal lastd : std_logic := '0';
Begin
     P1: Process ( clk )
     Begin
             If rising_edge(clk) then
                 if ( d = '0' ) then
                         Q <= '1';
                         Lastd <= '0';
                 elsif ( d = '1' ) then
                         Q <= '0';
                         Lastd <= '1';
```

```
                Else
                        Q <= 'X';
                        Lastd <= 'X';
                    End if;
            End if;
        If falling_edge(clk) then
                If ( lastd = '0' ) then
                        Q <= '0';
                elsif ( lastd = '1' ) then
                        Q <= '1';
        Else
                        Q <= 'X';
            End if
        End if;
    End process p1;
 End architecture basic;
```

This VHDL is simple but there is an even simpler way to encode the data and that is to simply XOR the clock with the data. If we look at the same data sequence, we can see that if we add a clock, and observe the original data and the Manchester encoded output, that this is simple the data XORd with the clock shown in Figure 17.

So, using this simple mechanism, we can create a much simpler Manchester encoder that simply XORs the clock and the data to obtain the resulting Manchester encoded data stream with the resulting VHDL:

```
Library ieee;
Use ieee.std_logic_1664.all;

Entity Manchester_encoder is
    Port (
            Clk : in std_logic;
            D : in std_logic;
            Q : out std_logic
        );
End entity Manchester_encoder;

Architecture basic of Manchester_encoder is
Begin
        Q <= D XOR CLK;
End architecture basic;
```

Decoding the Manchester data stream is also a choice between asynchronous and synchronous approaches. We can use a local clk and detect the values of the input to evaluate whether the values on the rising and falling edges are 0 or 1, respectively and ascertain the values of the data as a result, but clearly this is dependent on the

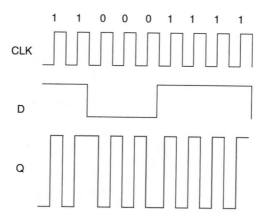

Figure 17
Manchester Encoding
Using XOR Function

transmitter and receiver clocks being synchronized to a reasonable degree. Such a simple decoder could look like this:

```
Entity Manchester_decoder is
      Port (
              Clk : in std_logic;
              D : in std_logic;
              Q : out std_logic
      );
End entity Manchester_decoder;

Architecture basic of Manchester_decoder is
      Signal lastd : std_logic := '0';
Begin
      P1 : process (clk)
      Begin
              If rising_edge(clk) then
                   Lastd <= d;
              End if;
      If falling_edge(clk) then
              If (lastd = '0') and (d = '1') then
                   Q <= '1';
              Elsif (lastd = '1') and (d = '0') then
                   Q <= '0';
              Else
                   Q <= 'X';
              End if;
      End if;
   End process p1;
End architecture basic;
```

In this VHDL model, the clock is at the same rate as the transmitter clock, and the data should be sent in packets (see the data checking chapter of this book) to ensure that the data is not sent in too large blocks such that the clock can get out of sync, and also that the data can be checked for integrity to correct for mistakes or the clock on the receiver being out of phase.

NRZ coding and decoding

The NRZ encoding scheme is actually not a coding scheme at all. It simply states that a '0' is transmitted as a '0' and a '1' is transmitted as a '1'. It is only worth mentioning because a designer may see the term NRZ and assume that a specific encoder or decoder was required, whereas in fact this is not the case. It is also worth noting that there are some significant disadvantages in using this simple approach. The first disadvantage, especially when compared to the Manchester coding scheme is that long sequences of '0's or '1's give effectively DC values when transmitted, that are susceptible to problems of noise and also make clock recovery very difficult. The other issue is that of bandwidth. Again if we compare the coding scheme to that of the Manchester example, it is obvious that the Manchester scheme requires quite a narrow bandwidth (relatively) to transmit the data, whereas the NRZ scheme may require anything from DC up to half the data rate (Nyquist frequency) and anything in between. This makes line design and filter design very much more problematic.

NRZI coding and decoding

In the NRZI scheme, the potential problems of the NRZ scheme, particularly the long periods of DC levels are partially alleviated. In the NRZI, if the data is a '0', then the data does not change, whereas if a '1' occurs on the data line, then the output changes. Therefore the issue of long sequences of '1's is addressed, but the potential for long sequences of '0's remains. It is a simple matter to create a basic model for a NRZI encoder using the following VHDL model:

```
Entity nrzi_encoder is
    Port (
            CLK : in std_logic;
            D : in std_logic;
            Q : out std_logic
        );
End entity nrzi_encoder;

Architecture basic of nrzi_encoder is
    Signal qint : std_logic := '0';
```

```
Begin
     p1 : process (clk)
     Begin
          If (d = '1') then
               If ( qint = '0' ) then
                    Qint <= '1';
               else
                    Qint <= '0';
               End if;
          End if;
     End process p1;
     Q <= qint;
End architecture basic;
```

Notice that this model is synchronous, but if we wished to make it asynchronous, the only changes would be to remove the clk port and change the process sensitivity list from clk to d. We can apply the same logic to the output, to obtain the decoded data stream, using the VHDL below. Again we are using a synchronous approach:

```
Entity nrzi_decoder is
     Port (
          CLK : in std_logic;
          D : in std_logic;
          Q : out std_logic
     );
End entity nrzi_decoder;

Architecture basic of nrzi_decoder is
     Signal lastd : std_logic := '0';
Begin
     p1 : process (clk)
     Begin
          If rising_edge(clk) then
               If (d = lastd) then
                    Q <= '0';
               Else
                    Q <= '1';
               End if;
               Lastd <= d;
          End if;
     End process p1;
End architecture basic;
```

The NRZI decoder is extremely simple, in that the only thing we need to check is whether the data stream has changed since the last clock edge. If the data has changed since the last clock, then we know that the data is a '1', but if the data is unchanged, then we know that it is a '0'. Clearly we could use an asynchronous approach, but this would rely on the data checking algorithm downstream being synchronized correctly.

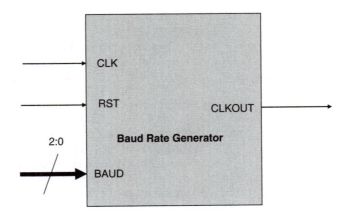

Figure 18
Baud Clock Generator

RS-232

Introduction

The basic approach of RS-232 serial transmission is that of a UART. UART stands for Universal Asynchronous Receiver/Transmitter. It is the standard method of translating a serial communication stream into the parallel form used by computers. RS-232 is a UART that has a specific standard defined for start, stop, break, data, parity and pin names.

RS-232 baud rate generator

The RS-232 is an asynchronous transmission scheme and so the correct clock rate must be defined prior to transmission to ensure that the data is transmitted and received correctly. The RS-232 baud rate can range from 1200 baud up to 115200 baud. This is based on a standard clock frequency of 14.7456 MHz, and this is then divided down by 8,16,28,48,96,192,384 and 768 to get the correct baud rates. We therefore need to define a clock divider circuit that can output the correct baud rate configured by a control word. We have obviously got 8 different ratios, and so we can use a 3 bit control word (baud[2:0]) plus a clock and reset to create the correct frequencies, assuming that the basic clock frequency is 14.7456 MHz (Figure 18).

The VHDL for this controller is given below and uses a single process to select the correct baud rate and another to divide down the input clock accordingly:

```
LIBRARY ieee;
USE ieee.Std_logic_1164.ALL;
USE ieee.Std_logic_unsigned.ALL;
```

```
ENTITY baudcontroller IS
    PORT(
            clk : IN std_logic;
            rst : IN std_logic;
            baud : IN std_logic_vector(0 to 2);
            clkout : OUT std_logic);
END baudcontroller;

ARCHITECTURE simple OF baudcontroller IS
    SIGNAL clkdiv : integer := 0;
        SIGNAL count : integer := 0;
BEGIN
Div: process (rst, clk)
begin
    if rst = '0' then
            clkdiv <= 0;
            count <= 0;
    elsif rising_edge(CLK) then
    case Baud is
            when "000" => clkdiv <= 7;   -- 115200
            when "001" => clkdiv <= 15;  -- 57600
            when "010" => clkdiv <= 23;  -- 38400
            when "011" => clkdiv <= 47;  -- 19200
            when "100" => clkdiv <= 95;  -- 9600
            when "101" => clkdiv <= 191; -- 4800
            when "110" => clkdiv <= 383; -- 2400
            when "111" => clkdiv <= 767; -- 1200
            when others => clkdiv <= 7;
    end case;
    end if;
end process;

clockdivision: process (clk, rst)
begin
    if rst='0' then
            clkdiv <= 0;
            count <= 0;
    elsif rising_edge(CLK) then
            count <= count + 1;
            if (count > clkdiv) then
                    clkout <= not clkout;
                    count <= 0;
            end if;
    end if;
end process;
END simple;
```

RS-232 receiver

The RS-232 receiver wait for data to arrive on the RX line and has a specification defined as follows <number of bits><parity>

Figure 19
Serial Data Receiver

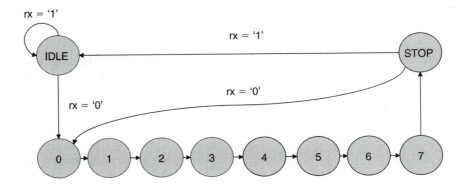

Figure 20
Basic Serial
Receiver

<stop bits>. So, for example an 8 bit, No parity, 1 stop bit specification would be given as 8N1. The RS-232 voltage levels are between $-12\,V$ and $+12\,V$, and so we will assume that an interface chip has translated these to standard logic levels (e.g. 0–5 V or 0–3.3 V). A sample bit stream would be of the format shown in Figure 19.

The idle state for RS-232 is high, and in this figure, after the stop bit, the line is shown as going low, in fact that only happens when another data word is coming. If the data transmission has finished, then the line will go high ('idle') again. We can in fact model this as a simple state machine as shown in Figure 20.

We can implement this simple state machine in VHDL using the following model:

```
LIBRARY ieee;
USE ieee.Std_logic_1164.ALL;
USE ieee.Std_logic_unsigned.ALL;

ENTITY serialrx IS
      PORT(
            clk : IN std_logic;
            rst : IN std_logic;
            rx : IN std_logic;
            dout : OUT std_logic_vector (7 downto 0)
            );
END serialrx;
```

```vhdl
ARCHITECTURE simple OF serialrx IS
   type state is (idle, s0, s1, s2, s3, s4, s5, s6, s7,
     stop);
   signal current_state, next_state : state;
   signal databuffer : std_logic_vector(7 downto 0);
BEGIN
receive: process (rst, clk)
begin
      if rst='0' then
                  current_state <= idle;
                  for i in 7 downto 0 loop
                        dout(i) <= '0';
                  end loop;
      elsif rising_edge(CLK) then


                  case current_state is
                  when idle =>
                  if rx = '0' then
                        next_state <= s0;
                  else
                        next_state <= idle;
                        end if;
                  when s0 =>
                        next_state <= s1;
                        databuffer(0) <= rx;
                  when s1 =>
                        next_state <= s2;
                        databuffer(1) <= rx;
                  when s2 =>
                        next_state <= s3;
                        databuffer(2) <= rx;
                  when s3 =>
                        next_state <= s4;
                        databuffer(3) <= rx;
                  when s4 =>
                        next_state <= s5;
                        databuffer(4) <= rx;
                  when s5 =>
                        next_state <= s6;
                        databuffer(5) <= rx;
                  when s6 =>
                        next_state <= s7;
                        databuffer(6) <= rx;
                  when s7 =>
                        next_state <= stop;
                        databuffer(7) <= rx;
                  when stop =>
                        if rx = '0' then
                              next_state <= s0;
                        else
                              next_state <= idle;
                              end if;
```

```
                    dout <= databuffer;
            end case;
            current_state <= next_state;
        end if;
    end process;
END;
```

Universal Serial Bus

The USB protocol has become pervasive and ubiquitous in the computing and electronics industries in recent years. The protocol supports a variety of data rates from low speed (10–100 kbits/s) up to high speed devices (up to 400 Mbits/s). While in principle it is possible to create Field Programmable Gate Array (FPGA) interfaces directly to a USB bus, for anything other than the lower data rates it requires accurate voltage matching and impedance matching of the serial bus. For example, the low data rates require 2.8 V ('1') and 0.3 V ('0'), differentially, whereas the high speed bus requires 400 mV signals, and in both cases termination resistors are required.

In practice, therefore, it is common when working with FPGAs to use a simple interface chip that handles all the analogue interface issues and can then be connected directly to the FPGA with a simple UART style interface. An example device is the Silicon Labs CP2101, that takes the basic USB Connector pins (Differential Data and Power and Ground) and then sets up the basic serial data transmission pins. The block diagram of this device is given in Figure 21.

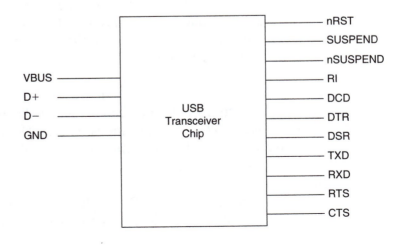

Figure 21
USB Transceiver
Chip CP2101

The pins on this device are relatively self explanatory and are summarized below:

nRST	The Reset pin for the device – Active Low
Suspend	This pin shows when the USB device is in SUSPEND mode – Active High
nSuspend	The Active Low (i.e. inverse) of the SUSPEND pin
RI	Ring Indicator
DCD	Data Carrier Detection – shows that data is on the USB line – Active low
DTR	Data Transmit Detection – this is Active Low when the line is ready for data transmission
DSR	Digital Sound Reconstruction
TXD	Asynchronous Data transmission line
RXD	Asynchronous Data received line
RTS	Clear to Receive – Active Low
CTS	Clear to Send – Active Low

The basic operation of the serial port starts from the use of the TXD and RXD (data) lines. If the configuration is as a NULL modem with no handshaking, it is possible to simply use the transmit (TXD) and receive (RXD) lines alone.

If you wish to check that the line is clear for sending data, then the RTS signal can be set (Request to Send), in this case Active Low, and if the line is ready, then the CTS line will go low and the data can be sent. This basic scheme is defined in such a way that once the receiver signal goes low, that the transmitter can send at any rate, the assumption being that the receiver can handle whatever rate is provided.

The protocol can be made more capable by using the DTR line, and this notifies the other end of the link that the device is ready for receiving data communications. The DCD line is not used directly in the link, but indicates that there is a valid communications link between the devices.

We can develop a VHDL model for such a communications link with as much complexity as we need to communicate with the hardware in the system under consideration, starting with a simple template:

```
Entity serial_handler is
        Port(
                Clk : in std_logic;
                Nrst : in std_logic;
                Data_in : in std_logic;
                Data_out : out std_logic;
                TXD : out std_logic;
                RXD : in std_logic
        );
End entity serial_handler;
```

In this initial model, we have a simple clock and reset, with two data connections for the synchronous side, and the TXD and RXD asynchronous data communications lines. We can put together a simple architecture that simply samples the data lines and transfers them into an intermediate variable for use on the synchronous side of the model:

```
Architecture basic of serial_handler is
Begin
        p1 : process (clk)
        Begin
                If rising_edge(clk) then
                        Rxd_int <= rxd;
                End if;
        End process p1;
End architecture basic;
```

We can extend this model to handle the transmit side also, using a similar approach:

```
Architecture basic of serial_handler is
Begin
        p1 : process (clk)
        Begin
                If rising_edge(clk) then
                        Data_out <= rxd;
                        Txd <= data_in;
                End if;
        End process p1;
End architecture basic;
```

This entity is the equivalent to a NULL modem architecture. If we wish to add the DTR notification that the device is ready for receiving data, we can add this to the entity list of ports and then gate the receive data if statement using the DTR signal:

```
Entity serial_handler is
     Port (
             Clk : in std_logic;
             Nrst : in std_logic;
             Data_in : in std_logic;
             Data_out : out std_logic;
             DTR : in std_logic;
             TXD : out std_logic;
             RXD : in std_logic
     );
End entity serial_handler;
Architecture serial_dtr of serial_handler is
Begin
     p1 : process (clk)
     Begin
             If rising_edge(clk) then
             If DTR = '0' then
                   Data_out <= rxd;
             End if;
             Txd <= data_in;
     End if;
   End process p1;
End architecture basic;
```

Using this type of approach we can extend the serial handler to incorporate as much or as little of the modem communications link protocol as we require.

Summary

In this chapter we have introduced a variety of serial communications coding and decoding schemes, and also reviewed the practical methods of interfacing using RS-232 and a USB device. Clearly, there are many more variations on this theme, and in fact a complete USB handler description would be worthy of a complete book in itself.

8
Digital Filters

Introduction

An important part of systems that interface to the 'real world' is the ability to process sampled data in the digital domain. This is often called Sampled Data Systems (SDS) or operating in the Z-domain. Most engineers are familiar with the operation of filters in the Laplace or S-domain where a continuous function defines the characteristics of the filter and this is the digital domain equivalent to that.

For example, consider a simple RC circuit in the analog domain as shown in Figure 22. This is a low pass filter function and can be represented using the Laplace notation shown in Figure 22.

This has the equivalent S-domain (or Laplace) function as follows:

$$L(s) = \frac{1}{1 + sCR}$$

This function is a low pass filter because the Laplace operator s is equivalent to $j\omega$, where $w = 2\pi f$ (with f being the frequency). If f is zero (the DC condition), then the gain will be 1, but if the value of sCR is equal to 1, then the gain will be 0.5. This in dB is $-3\,dB$ and is the classical low pass filter cut off frequency.

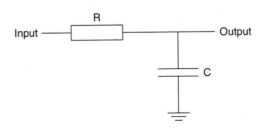

Figure 22
RC Filter in the
Analog Domain

In the digital domain, the s operation is replaced by Z. Z^{-1} is practically equivalent to a delay operator, and similar functions to the Laplace filter equations can be constructed for the digital, or Z-domain equivalent.

There are a number of design techniques, many beyond the scope of this book (if the reader requires a more detailed introduction to the realm of digital filters, Cunningham's *Digital filtering: an introduction* is a useful starting point), however it is useful to introduce some of the basic techniques used in practice and illustrate them with examples.

The remainder of this chapter will cover the introduction to the basic techniques and then demonstrate how these can be implemented using VHDL on Field Programmable Gate Array (FPGAs).

Converting S-domain to Z-domain

The method of converting an S-domain equation for a filter to its equivalent Z-domain expression is done using the 'bilinear transform'. This is simply a method of expressing the equation in the S-domain in terms of Z. The basic approach is to replace each instance of s with its equivalent Z-domain notation and then rearrange into the most convenient form. The transform is called bilinear as both the numerator and denominator of the expression are linear in terms of z.

$$s = \frac{z - 1}{z + 1}$$

If we take a simple example of a basic second order filter we can show how this is translated into the equivalent Z-domain form:

$$H(s) = \frac{1}{s^2 + 2s + 1}$$

replace s with $(z - 1)/(z + 1)$:

$$H(Z) = \frac{1}{\left(\frac{(z - 1)}{(z + 1)}\right)^2 + 2\frac{(z - 1)}{(z + 1)} + 1}$$

$$H(Z) = \frac{(z + 1)^2}{(z - 1)^2 + (z - 1)(z + 1) + (z + 1)^2}$$

$$H(Z) = \frac{z^2 + 2z + 1}{3z^2 + 1}$$

Now, the term $H(Z)$ is really the output $Y(Z)$ over the input $X(Z)$ and we can use this to express the Z-domain equation in terms of the input and output:

$$H(Z) = \frac{z^2 + 2z + 1}{3z^2 + 1}$$

$$\frac{Y(Z)}{X(Z)} = \frac{z^2 + 2z + 1}{3z^2 + 1}$$

$$3z^2 Y(Z) + Y(Z) = z^2 X(Z) + 2z X(Z) + X(Z)$$

This can then be turned into a sequence expression using delays (z is one delay, z^2 is two delays and so on) with the following result:

$$3z^2 Y(Z) + Y(Z) = z^2 X(Z) + 2z X(Z) + X(Z)$$

$$3y(n + 2) + y(n) = x(n+) + 2x(n + 1) + x(n)$$

This is useful because we are now expressing the Z-domain equation in terms of delay terms, and the final step is to express the value of $y(n)$ (the current output) in terms of past elements by reducing the delays accordingly (by 2 in this case):

$$3y(n + 2) + y(n) = x(n + 2) + 2x(n + 1) + x(n)$$

$$3y(n) + y(n - 2) = x(n) + 2x(n - 1) + x(n - 2)$$

$$y(n) + 1/3y(n - 2) = 1/3x(n) + 2/3x(n - 1) + 1/3x(n - 2)$$

$$y(n) = 1/3x(n) + 2/3x(n - 1) + 1/3x(n - 2) - 1/3y(n - 2)$$

The final design note at this point is to make sure that the design frequency is correct, for example the low pass cut off frequency. The frequencies are different between the S- and Z-domain models, even after the bilinear transformation, and in fact the desired digital domain frequency must be translated into the equivalent S-domain frequency using a technique called *pre-warping*. This simple step translates the frequency from one domain to the other using the expression below:

$$\omega_c = \tan\left(\frac{\Omega_c T}{2}\right)$$

Where Ω_c is the digital domain frequency, T is the sampling period of the Z-domain system and ω_c is the resulting frequency for the analog domain calculations.

Once we have obtained our Z-domain expressions, how do we turn this into practical designs? The next section will explain how this can be achieved.

Implementing Z-domain functions in VHDL

Introduction

Z-domain functions are essentially digital in the time domain as they are discrete and sampled. The functions are also discrete in the amplitude axis, as the variables or signals are defined using a fixed number of bits in a real hardware system, whether this is integer, signed, fixed point or floating point, there is always a finite resolution to the signals. For the remainder of this chapter, signed arithmetic is assumed for simplicity and ease of understanding. This also essentially defines the number of bits to be used in the system. If we have 8 bits, the resolution is 1 bit and the range is $-128-+127$.

Gain block

The first main Z-domain block is a simple gain block. This requires a single signed input, a single signed output and a parameter for the gain. This could be an integer or also a signed value. The VHDL model for a simple Z-domain gain block is given below:

```
Library ieee;
Use ieee.numeric_std.all;

Entity zgain is
      Generic ( n : integer := 8;
                  gain : signed
        );
        Port (
                Zin : in signed (n-1 downto 0);
                Zout : out signed (n-1 downto 0)
        );
End entity zgain;

Architecture zdomain of zgain is
Begin
      p1 : process(zin)
              variable product : signed (2*n-1 downto 0);
        begin
                product := zin * gain;
                zout <= product (n-1 downto 0);
        end process p1;
End architecture zdomain;
```

We can test this with a simple testbench that ramps up the input and we can observe the output being changed in turn:

```
library ieee;
use ieee.std_logic_1164.all;
use ieee.numeric_std.all;
```

```
entity tb is
end entity tb;

architecture testbench of tb is

        signal clk : std_logic := '0';
        signal dir : std_logic := '0';
        signal zin : signed (7 downto 0):= X"00";
        signal zout : signed (7 downto 0):= X"00";

        component zgain
        generic (
                n : integer := 8;
                gain :signed := X"02"
        );
        port (
                signal zin : in signed(n-1 downto 0);
                signal zout : out signed(n-1 downto 0)
        );
        end component;
        for all : zgain use entity work.zgain;

begin
        clk <= not clk after 1 us;

        DUT : zgain generic map ( 8, X"02" ) port map
          (zin, zout);

        p1 : process (clk)
        begin
                zin <= zin + 1;
        end process p1;
end architecture testbench;
```

Clearly, this model has no error checking or range checking and the obvious problem with this type of approach is that of overflow. For example, if we multiply the input (64) by a gain of 2, we will get 128, but that is the sign bit, and so the result will show -128! This is an obvious problem with this simplistic model and care must be taken to ensure that adequate checking takes place in the model.

Sum and difference

Using this same basic approach, we can create sum and difference models which are also essential building blocks for a Z-domain system. The sum model VHDL is shown below:

```
Library ieee;
Use ieee.numeric_std.all;
```

```
Entity zsum is
    Generic ( n : integer := 8
    );
    Port (
            Zin1 : in signed (n-1 downto 0);
            Zin2 : in signed (n-1 downto 0);
            Zout : out signed (n-1 downto 0)
    );
End entity zsum;

Architecture zdomain of zsum is
Begin
    p1 : process(zin)
        variable zsum : signed (2*n-1 downto 0);
    begin
        zsum := zin1 + zin2;
        zout <= zsum (n-1 downto 0);
    end process p1;
End architecture zdomain;
```

Despite the potential for problems with overflow, both of the models shown have the internal variable that is twice the number of bits required, and so this can take care of any possible overflow internal to the model, and in fact checking could take place prior to the final assignment of the output to ensure the data is correct. The difference model is almost identical to the sum model except that the difference of zin1 and zin2 is computed.

Division model

A useful model for scaling numbers simply in the Z-domain is the division by 2 model. This model simply shifts the current value in the input to the right by 1 bit – hence giving a division by 2. The model could easily be extended to shift right by any number of bits, but this simple version is very useful by itself. The VHDL for the model relies on the logical shift right operator (SRL) which not only shifts the bits right (losing the least significant bit) but adding a zero at the most significant bit. The resulting VHDL is shown below for this specific function:

```
zout <= zin srl 1;
```

The unit shift can be replaced by any integer number to give a shift of a specific number of bits. For example, to shift right by 3 bits (effectively a divide by 8) would have the following VHDL:

```
zout <= zin srl 3;
```

The complete division by 2 model is given below:

```
Library ieee;
Use ieee.numeric_std.all;

Entity zdiv2 is
        Generic ( n : integer := 8
        );
        Port (
                Zin : in signed (n-1 downto 0);
                Zout : out signed (n-1 downto 0)
        );
End entity zdiv2;

Architecture zdomain of zdiv2 is
Begin
        zout <= zin srl 1;
End architecture zdomain;
```

In order to test the model a simple test circuit that ramps up the input is used and this is given below:

```
library ieee;
use ieee.std_logic_1164.all;
use ieee.numeric_std.all;

entity tb is
end entity tb;

architecture testbench of tb is

        signal clk : std_logic := '0';
        signal dir : std_logic := '0';
        signal zin : signed (7 downto 0) := X"00";
        signal zout : signed (7 downto 0) := X"00";

        component zdiv2
        generic (
                n : integer := 8
        );
        port (
                signal zin : in signed (n-1 downto 0);
                signal zout : out signed (n-1 downto 0)
        );
        end component;
        for all : zdiv2 use entity work.zdiv2;

    begin
        clk <= not clk after 1 us;

        DUT : zdiv2 generic map (8) port map (zin, zout);

        p1 : process (clk)
```

```
        begin
                zin <= zin + 1;
        end process p1;
end architecture testbench;
```

The behavior of the model is useful to review, if the input is X"03" (Decimal 3), binary '00000011' and the number is right shifted by 1, then the resulting binary number will be '00000001' (X"01" or decimal 1), in other words this operation always rounds down. This has obvious implications for potential loss of accuracy and the operation is skewed downwards, which has again, implications for how numbers will be treated using this operator in a more complex circuit.

Unit delay model

The final basic model is the unit delay model (zdelay). This has a clock input (clk) using a std_logic signal to make it simple to interface to standard digital controls. The output is simply a one clock cycle delayed version of the input.

```
Library ieee;
use ieee.std_logic_1164.all;
Use ieee.numeric_std.all;

Entity zdelay is
        Generic ( n : integer := 8 );
        Port (
                clk : in std_logic;
                Zin : in signed (n-1 downto 0);
                Zout : out signed (n-1 downto 0) := (others
                   => '0')
        );
End entity zdelay;

Architecture zdomain of zdelay is
        signal lastzin : signed (n-1 downto 0) := (others
           => '0');
Begin
        p1 : process(clk)
        begin
                if rising_edge(clk) then
                        zout <= lastzin;
                        lastzin <= zin;
                end if;
        end process p1;
End architecture zdomain;
```

Notice that the output zout is initialized to all zeros for the initial state, otherwise 'don't care' conditions can result that propagate across the complete model.

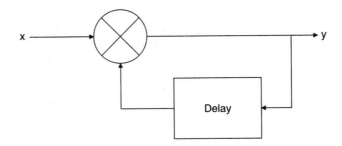

Figure 23
Simple Z-Domain Low
Pass Filter

Basic low pass filter model

We can put these elements together in simple models that implement basic filter blocks in any configuration we require, as always taking care to ensure that overflow errors are checked for in practice.

To demonstrate this, we can implement a simple low pass filter using the basic block diagram shown in Figure 23.

We can create a simple test circuit that uses the individual models we have already shown for the sum and delay blocks and apply a step change and observe the response of the filter to this stimulus. (Clearly, in this case, with unity gain the filter exhibits positive feedback and so to ensure the correct behavior we use the divide by 2 model zdiv2 in both the inputs to the sum block to ensure gain of 0.5 on both. These are not shown in the Figure 23.) The resulting VHDL model is shown below (note the use of the zdiv2 model):

```
library ieee;
use ieee.std_logic_1164.all;
use ieee.numeric_std.all;

entity tb is
end entity tb;

architecture testbench of tb is
        signal clk : std_logic := '0';
        signal x : signed (7 downto 0):= X"00";
        signal y : signed (7 downto 0):= X"00";
        signal y1 : signed (7 downto 0):= X"00";
        signal yd : signed (7 downto 0):= X"00";
        signal yd2 : signed (7 downto 0):= X"00";
        signal x2 : signed (7 downto 0):= X"00";
component zsum
generic (
        n : integer : = 8
);
```

```
port (
        signal zin1 : in signed(n-1 downto 0);
        signal zin2 : in signed(n-1 downto 0);
        signal zout : out signed(n-1 downto 0)
);
end component;
for all : zsum use entity work.zsum;

component zdiff
generic (
        n : integer := 8
);
port (
        signal zin1 : in signed(n-1 downto 0);
        signal zin2 : in signed(n-1 downto 0);
        signal zout : out signed(n-1 downto 0)
);
end component;
for all : zdiff use entity work.zdiff;

    component zdiv2
    generic (
            n : integer := 8
    );
    port (
            signal zin : in signed(n-1 downto 0);
            signal zout : out signed(n-1 downto 0)
    );
    end component;
    for all : zdiv2 use entity work.zdiv2;

    component zdelay
    generic (
            n : integer := 8
    );
    port (
            signal clk : in std_logic;
            signal zin : in signed(n-1 downto 0);
            signal zout : out signed(n-1 downto 0)
    );
    end component;
    for all : zdelay use entity work.zdelay;

begin
    clk <= not clk after 1 us;

    GAIN1 : zdiv2 generic map (8) port map (x, x2);
    GAIN2 : zdiv2 generic map (8) port map (yd, yd2);
    SUM1 : zsum generic map (8) port map (x2, yd2,
      y);
    D1 : zdelay generic map (8) port map (clk, y,
      yd);

    x <= X"00", X"0F" after 10 us;
end architecture testbench;
```

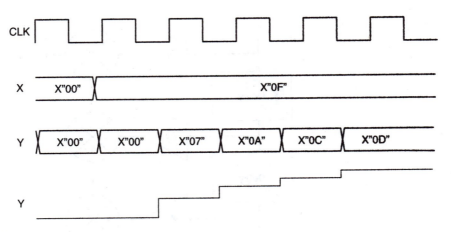

Figure 24
Basic Low Pass Filter Simulation Waveforms

The test circuit applies a step change of X"00" to X"0F" after 10 us, and this results in the filter response. We can show this graphically in Figure 24 with the output in both Hexadecimal and 'analog' form for illustration.

It is interesting to note the effect of using the zdiv2 function on the results. With the input of 0F (binary 00001111) we lose the LSB when we divide by 2, giving the resulting input to the sum block of 00000111 (7) which added together with the division of the output gives a total of 14 as the maximum possible output from the filter. In fact, the filter gives an output of X"0D" or binary 00001101, which is two down from the theoretical maximum of X"0F" and this highlights the practical difficulties when using a 'coarse' approximation technique for numerical work rather than a fixed or floating point method. On the other hand, it is clearly a simple and effective method of implementing a basic filter in VHDL.

Later in this book, the use of fixed and floating point numbers are discussed, as is the use of multiplication for more exact calculations and for practical filter design, where higher accuracy is required, then it is likely that both these methods would be used. There may be situations, however, where it is simply not possible to use these advanced techniques, particularly a problem when space is at a premium on the FPGA and in these cases, the simple approach described in this chapter will be required.

There are numerous texts on more advanced topics in digital filter design, and these are beyond the scope of this book, but it is

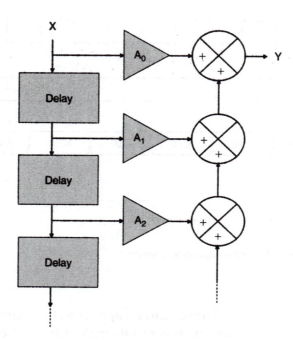

Figure 25
FIR Filter Schematic

useful to introduce some key concepts at this stage of the two main types of digital filter in common usage today. These are the recursive (or Infinite Impulse Response – IIR) filters and non-recursive (or Finite Impulse Response – FIR) filters.

FIR filters

FIR filters are characterized by the fact that they use only delayed versions of the input signal to filter the input to the output. For example, if we take the expression for a general FIR filter below, we can see that the output is a function of a series of delayed, scaled versions of the input:

$$y = \sum_{i=0}^{n} A_i x[i]$$

Where A_i is the scale factor for the ith delayed version of the input. We can represent this graphically in the diagram shown in Figure 25. We can implement this model using the basic building blocks described in this chapter of gain, division, sums and delays to develop block based models for such filters. As noted in the previous section, it is important to ensure that for higher accuracy filters, fixed or floating point arithmetic is required and also the use of multipliers for added accuracy is preferable in most cases to that of simple gain and division blocks as described previously in this chapter.

IIR filters

IIR filters are characterized by the fact that they use delayed versions of the input signal and fed back and delayed version of the output signal to filter the input to the output. For example, if we take the expression for a general IIR filter below, we can see that the output is a function of a series of delayed, scaled versions of the input and output:

$$y = \sum_{i=0}^{n} \frac{A_i x[i]}{B_i y[i]}$$

Where A_i is the scale factor for the ith delayed version of the input and B_i is the scale factor for the ith delayed version of the output. This is obviously very similar to the FIR example previously and can be built up using the same basic elements. If we consider the simple example earlier in this chapter, it can be seen that this is in fact a simple first order (single delay) IIR filter, with no delayed versions of the input and a single delayed version of the output.

Summary

This chapter has introduced the concepts of implementing basic digital filters in VHDL and has given examples of both the building blocks and constructed filters for implementation on an FPGA platform. The general concepts of FIR and IIR filters have been introduced so that the reader can implement the topology and type of filter appropriate for their own application.

9

Secure Systems

Introduction to block ciphers

The Data Encryption Standard (DES) is a symmetric 'block cipher'. A stream cipher operates on a digital data stream one or more bits at a time. A block cipher operates on complete blocks of data at any one time and produces a ciphertext block of equal size. DES is a block cipher that operates on data blocks of 64 bits in size. DES uses a 64-bit key – 8×8 including 1 bit for parity, so the actual key is 56 bits. DES, in common with other block ciphers is based around a structure called a 'FEISTEL LATTICE', so it is useful to describe how this works.

Feistel lattice structures

A block cipher operates on a plaintext block of n bits to produce a block of ciphertext of n bits. For the algorithm to be reversible (i.e. for decryption to be possible) there must be a unique mapping between the two sets of blocks. This can also be called a non-singular transformation. For example, consider the transformations given in Figure 26.

Obviously this is essentially a substitution cipher, that may be susceptible to the standard statistical analysis techniques used for

Figure 26
Reversible and
Irreversible
Transformations

Reversible		Irreversible	
Plaintext	Ciphertext	Plaintext	Ciphertext
00	11	00	11
01	10	01	10
10	00	10	10
11	10	11	10

simple cryptanalysis of text (such as frequency analysis). As the block size increases, then this becomes increasingly less feasible. An obvious practical difficulty with this approach is the number of transformations required as n increases. This mapping is essentially the key and for an n bit general substitution block cipher, the key size is $n \times 2^n$. For $n = 64$, the key size becomes $64 \times 2^{64} \sim= 10^{21}$

In order to get around this complexity problem, Feistel proposed an approach called a 'product cipher' whereby the combination of several simple steps leads to a much more cryptographically secure solution than any of the component ciphers used. His approach relies on the alternation of two types of function:

- Diffusion

- Confusion

These two concepts are grounded in an approach developed by Shannon and is used in most standard block ciphers in common use today. Shannon's goal was to define cryptographic functions that would not be susceptible to statistical analysis. Shannon proposed two methods for reducing the ability of statistical cryptanalysis to find the original message – diffusion and confusion.

In diffusion, the statistical structure of the plaintext is dissipated throughout the long term statistics of the ciphertext. This is achieved by making each bit of the plaintext affect the value of many bits of the ciphertext. An example of this would be to add letters to a ciphertext such that the frequency of each letter is the same – regardless of the message. In binary block ciphers the technique uses multiple permutations and functions such that each bit of the ciphertext is affected by multiple bits in the plaintext.

Each block of plaintext is transformed into a block of ciphertext, and this depends on the key. Confusion aims to make the relationship between the ciphertext and the key as complex as possible to reduce the possibility of ascertaining the key. This requires a complex substitution algorithm as a linear substitution would not protect the key.

Both diffusion and confusion are the cornerstones of successful block cipher design.

The result of these requirements is the Feistel Lattice (shown in Figure 27). This is the basic architecture in ciphers such as DES.

The inputs to the algorithm are the plaintext (of length 2w bits) and a key K. The plaintext is split into two halves L and R, and the data

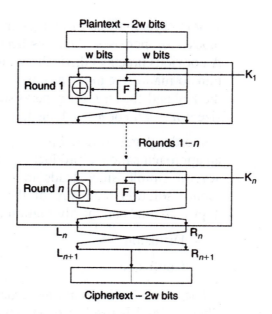

Figure 27
Feistel Lattice
Structure

is then passed through n 'rounds' of processing and then recombined to pro-duce the ciphertext. Each round has an input L_i-1 and R_i-1 derived from the previous round and a subkey K_i, derived from the overall key K. Each round has the same structure. The left halve of the data has a substitution performed. This requires a 'round func-tion' F to be performed on the right half of the data and then XORd with the left half. Finally a permutation is performed that requires the interchange of the two halves of the data.

The implementation of a Feistel network has the following key parameters:

- *Block size*: A larger block size generally means greater secu-rity, but reduced speed. 64-bit block sizes are very heavily used as being a reasonable trade-off although Advanced Encryption Standards (AES) now uses 128 bits.

- *Key Size*: The same trade-off applies as for block size. Generally 64 bits is not now considered adequate and 128 bits is preferred.

- *Number of rounds*: Each round adds additional security. A single round is inadequate, but 16 is considered standard.

- *Subkey generation*: The more complex this algorithm is, the more secure the overall system will be.

- *Round function*: Greater complexity again means greater resistance to cryptanalysis.

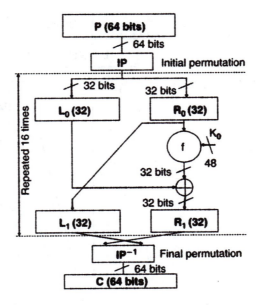

Figure 28
DES Coarse
Structure

The Data Encryption Standard

Introduction

The DES was adopted by the National Institute of Standards and Technology (NIST) in 1977 as the Federal Information Processing Standards 46 (FIPS PUB 46).

As mentioned previously, the algorithm operates on plaintext blocks of 64 bits and the key size is 56 bits. By 1999, NIST had decreed that DES was no longer secure and should only be used for legacy systems and that triple DES should be used instead. As will be described later, DES has since been superceded by the AES.

The coarse structure (overall architecture) of DES is shown in Figure 28.

The center section (where the main repetition occurs) is called the fine structure and is where the details of the encryption take place. This fine structure is detailed in Figure 29.

The fine structure of DES consists of several important functional blocks:

- *Initial permutation*: Fixed, known mapping 64×64 bits

- *Key transformations*: Circular L shift of keys by Ai bits in round (A(i) is known and fixed)

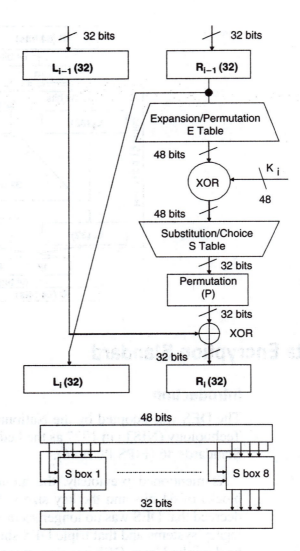

Figure 29
DES Fine Structure

Figure 30
S Box Architecture

- *Compression Permutation*: Fixed known subset of 56 bit input mapped onto 48 bit output

- *Expansion permutation*: 32 bit data shuffled and mapped (both operations fixed and known) onto 48 bits by duplicating 16 input bits. This makes diffusion quicker.

Another significant section of the algorithm is the substitution or S box. The Non-linear aspect of the cipher is vital in cryptography. In DES the 8 S boxes each contain 4 different (fixed & known) 4:4 input maps. These are selected by the extra bits created in the expansion box. The S boxes are structured as shown in Figure 30.

The final part of the DES structure is the key generation architecture for the individual round keys and this is given in Figure 31.

The remaining functional block is the initial and final permutation. The initial permutation (P box) is a 32:32 fixed, known bit permutation. The Final Permutation is the inverse of the initial permutation. The initial permutation is defined using the following table:

58	50	42	34	26	18	10	2
60	52	44	36	28	20	12	4
62	54	46	38	30	22	14	6
64	56	48	40	32	24	16	8
57	49	41	33	25	17	9	1
59	51	43	35	27	19	11	3
61	53	45	37	29	21	13	5
63	55	47	39	31	23	15	7

The final permutation is simply the inverse of the initial permutation.

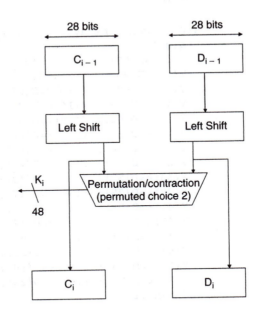

Figure 31
DES Round Key
Generation

DES VHDL implementation

DES can be implemented in VHDL using a structural or a functional approach. As has been discussed previously, there are advantages to both methods, however the DES algorithm is tied implicitly to the structure, so a structural approach will give an efficient implementation.

Implementing the initial permutation in VHDL requires a 64-bit input vector and a 64-bit output vector. We can create this in VHDL with an entity that defines an input and output std_logic vector as follows:

```
library ieee;
use ieee.std_logic_1164.all;

entity des_ip is port
        (
            D : in std_logic_vector(1 TO 64);
            Y : out std_logic_vector(1 TO 64)
        );
end des_ip;
```

The architecture is simply the assignment of bits from input to output according to the initial permutation table previously defined:

```
architecture behavior of des_ip is
begin
  Y(1)<=D(58);  Y(2)<=D(50);  Y(3)<=D(42);  Y(4)<=D(34);
  Y(5)<=D(26);  Y(6)<=D(18);  Y(7)<=D(10);  Y(8)<=D(2);
  Y(9)<=D(60);  Y(10)<=D(52); Y(11)<=D(44); Y(12)<=D(36);
  Y(13)<=D(28); Y(14)<=D(20); Y(15)<=D(12); Y(16)<=D(4);
  Y(17)<=D(62); Y(18)<=D(54); Y(19)<=D(46); Y(20)<=D(38);
  Y(21)<=D(30); Y(22)<=D(22); Y(23)<=D(14); Y(24)<=D(6);
  Y(25)<=D(64); Y(26)<=D(56); Y(27)<=D(48); Y(28)<=D(40);
  Y(29)<=D(32); Y(30)<=D(24); Y(31)<=D(16); Y(32)<=D(8);
  Y(33)<=D(57); Y(34)<=D(49); Y(35)<=D(41); Y(36)<=D(33);
  Y(37)<=D(25); Y(38)<=D(17); Y(39)<=D(9);  Y(40)<=D(1);
  Y(41)<=D(59); Y(42)<=D(51); Y(43)<=D(43); Y(44)<=D(35);
  Y(45)<=D(27); Y(46)<=D(19); Y(47)<=D(11); Y(48)<=D(3);
  Y(49)<=D(61); Y(50)<=D(53); Y(51)<=D(45); Y(52)<=D(37);
  Y(53)<=D(29); Y(54)<=D(21); Y(55)<=D(13); Y(56)<=D(5);
  Y(57)<=D(63); Y(58)<=D(55); Y(59)<=D(47); Y(60)<=D(39);
  Y(61)<=D(31); Y(62)<=D(23); Y(63)<=D(15); Y(64)<=D(7);
end behavior;
```

As this function is purely combinatorial we don't need to have a register (i.e. clocked input) on this model, although we could implement that if required using a simple process.

As shown in the previous description of the expansion function, we need to take a word consisting of 32 bits and expand it to 48 bits. This requires a translation table as shown below. Notice that there are duplicates in the cell which mean that you only need 32 input bits to obtain 48 output bits:

32	1	2	3	4	5
4	5	6	7	8	9
8	9	10	11	12	13
12	13	14	15	16	17
16	17	18	19	20	21
20	21	22	23	24	25
24	25	26	27	28	29
28	29	30	31	32	1

We can use a VHDL model similar to the initial permutation function, except that in this case there are 32 input bits and 48 output bits. Notice that some of the input bits are repeated giving a straight-forward expansion function:

```
library ieee;
use ieee.std_logic_1164.all;

entity des_e is port
        (
                D : in std_logic_vector(1 TO 32);
                Y : out std_logic_vector(1 TO 48)
        );
end des_e;
```

The architecture is simply the assignment of bits from input to output according to the initial permutation table previously defined:

```
architecture behavior of des_e is
begin
  Y(1)<=D(32);  Y(2)<=D(1);   Y(3)<=D(2);   Y(4)<=D(3);
  Y(5)<=D(4);   Y(6)<=D(5);   Y(7)<=D(4);   Y(8)<=D(5);
  Y(9)<=D(6);   Y(10)<=D(7);  Y(11)<=D(8);  Y(12)<=D(9);
  Y(13)<=D(8);  Y(14)<=D(9);  Y(15)<=D(10); Y(16)<=D(11);
  Y(17)<=D(12); Y(18)<=D(13); Y(19)<=D(12); Y(20)<=D(13);
  Y(21)<=D(14); Y(22)<=D(15); Y(23)<=D(16); Y(24)<=D(17);
  Y(25)<=D(16); Y(26)<=D(17); Y(27)<=D(18); Y(28)<=D(19);
  Y(29)<=D(20); Y(30)<=D(21); Y(31)<=D(20); Y(32)<=D(21);
  Y(33)<=D(22); Y(34)<=D(23); Y(35)<=D(24); Y(36)<=D(25);
  Y(37)<=D(24); Y(38)<=D(25); Y(39)<=D(26); Y(40)<=D(27);
  Y(41)<=D(28); Y(42)<=D(29); Y(43)<=D(28); Y(44)<=D(29);
  Y(45)<=D(30); Y(46)<=D(31); Y(47)<=D(32); Y(48)<=D(1);
end behavior;
```

The final 'permutation' block is the permutation marked (P) on the fine structure in Figure 29 after the key function. This is a straightforward bit substitution function with 32 bits input and 32 bits output. The bit translation table is shown in the table below:

16	7	20	21
29	12	28	17
1	15	23	26
5	18	31	10
2	8	24	14
32	27	3	9
19	13	30	6
22	11	4	25

This is implemented in VHDL using exactly the same approach as the previous expansion and permutation functions as follows:

```
library ieee;
use ieee.std_logic_1164.all;

entity des_p is port
        (
            D : in std_logic_vector(1 TO 32);
            Y : out std_logic_vector(1 TO 32)
        );
end des_p;
```

The architecture is simply the assignment of bits from input to output according to the initial permutation table previously defined:

```
architecture behavior of des_p is
begin
    Y(1)<=D(16);   Y(2)<=D(7);    Y(3)<=D(20);   Y(4)<=D(21);
    Y(5)<=D(29);   Y(6)<=D(12);   Y(7)<=D(28);   Y(8)<=D(17);
    Y(9)<=D(1);    Y(10)<=D(15);  Y(11)<=D(23);  Y(12)<=D(26);
    Y(13)<=D(5);   Y(14)<=D(18);  Y(15)<=D(31);  Y(16)<=D(10);
    Y(17)<=D(2);   Y(18)<=D(8);   Y(19)<=D(24);  Y(20)<=D(14);
    Y(21)<=D(32);  Y(22)<=D(27);  Y(23)<=D(3);   Y(24)<=D(9);
    Y(25)<=D(19);  Y(26)<=D(13);  Y(27)<=D(30);  Y(28)<=D(6);
    Y(29)<=D(22);  Y(30)<=D(11);  Y(31)<=D(4);   Y(32)<=D(25);
end behavior;
```

The non-linear part of the DES algorithm is the S box. This is a set of $6 \rightarrow 4$ bit transformations that reduce the 48 bits of the expanded word in the DES f function, to the 32 bits for the next round. The required row and column are obtained from the data passed into the S box. The data into the S box is a 6 bit binary word. The row is obtained from $2b_1 + b_6$ and the column is obtained from $b_2b_3b_4b_5$. For example, S(011011) would give a row of 01 (1) and a column of 1101 (13). For S8 this would result in a value returning of 1110 (14).

The basic S box entity can therefore be constructed using the following VHDL:

```
Library ieee;
Use ieee.std_logic_1164.all;
Entity des_sbox is
        Port (
                D : in std_logic_vector (1 to 6);
                Y : out std_logic_vector (1 to 4)
        );
End entity des_sbox;
```

One approach is to define the row and column from the input D word and then calculate the output Y word from that using a look up table approach or minimize the logic as a truth table. The basic architecture could then look something like this:

```
Architecture behaviour of sbox is
        Signal r : std_logic_vector (1 to 2);
        Signal c : std_logic_vector (3 to 6);
Begin
        R <= d ( 1 to 2);
        C <= d (3 to 6 );
        -- The look up table or logic goes here
End;
```

Another approach is to define a simple lookup table with the input D as the unique address and the output Y stored in the memory – this is exactly the same as a Read Only Memory (ROM), so the input is defined as an unsigned integer to look up the required value. In this case the memory is defined in exactly the same way as the ROM separately in this book.

The S box substitutions are given in the table below and the VHDL can either use the lookup table approach to store the address of each substitution, or logic can be used to decode the correct output.

In order to use this table, the appropriate S box is selected and then the two bits of the row select the appropriate row and the same for the column. For example, for S box S1, if the row is 3 (11) and the column is 10 (1010) then the output can be read off as 3 (0011). This can be coded in VHDL using nested case statements as follows:

```
Case row is
        When 0 =>
                Case column is
                        When 0 => y <= 14;
                        When 1 => y <= 4;
                        ...
                End case;
```

row	Column number															
	[0]	[1]	[2]	[3]	[4]	[5]	[6]	[7]	[8]	[9]	[10]	[11]	[12]	[13]	[14]	[15]
S_1																
[0]	14	4	13	1	2	15	11	8	3	10	6	12	5	9	0	7
[1]	0	15	7	4	14	2	13	1	10	6	12	11	9	5	3	8
[2]	4	1	14	8	13	6	2	11	15	12	9	7	3	10	5	0
[3]	15	12	8	2	4	9	1	7	5	11	3	14	10	0	6	13
S_2																
[0]	15	1	8	14	6	11	3	4	9	7	2	13	12	0	5	10
[1]	3	13	4	7	15	2	8	14	12	0	1	10	6	9	11	5
[2]	0	14	7	11	10	4	13	1	5	8	12	6	9	3	2	15
[3]	13	8	10	1	3	15	4	2	11	6	7	12	0	5	14	9
S_3																
[0]	10	0	9	14	6	3	15	5	1	13	12	7	11	4	2	8
[1]	13	7	0	9	3	4	6	10	2	8	5	14	12	11	15	1
[2]	13	6	4	9	8	15	3	0	11	1	2	12	5	10	14	7
[3]	1	10	13	0	6	9	8	7	4	15	14	3	11	5	2	12
S_4																
[0]	7	13	14	3	0	6	9	10	1	2	8	5	11	12	4	15
[1]	13	8	11	5	6	15	0	3	4	7	2	12	1	10	14	9
[2]	10	6	9	0	12	11	7	13	15	1	3	14	5	2	8	4
[3]	3	15	0	6	10	1	13	8	9	4	5	11	12	7	2	14
S_5																
[0]	2	12	4	1	7	10	11	6	8	5	3	15	13	0	14	9
[1]	14	11	2	12	4	7	13	1	5	0	15	10	3	9	8	6
[2]	4	2	1	11	10	13	7	8	15	9	12	5	6	3	0	14
[3]	11	8	12	7	1	14	2	13	6	15	0	9	10	4	5	3
S_6																
[0]	12	1	10	15	9	2	6	8	0	13	3	4	14	7	5	11
[1]	10	15	4	2	7	12	9	5	6	1	13	14	0	11	3	8
[2]	9	14	15	5	2	8	12	3	7	0	4	10	1	13	11	6
[3]	4	3	2	12	9	5	15	10	11	14	1	7	6	0	8	13
S_7																
[0]	4	11	2	14	15	0	8	13	3	12	9	7	5	10	6	1
[1]	13	0	11	7	4	9	1	10	14	3	5	12	2	15	8	6
[2]	1	4	11	13	12	3	7	14	10	15	6	8	0	5	9	2
[3]	6	11	13	8	1	4	10	7	9	5	0	15	14	2	3	12
S_8																
[0]	13	2	8	4	6	15	11	1	10	9	3	14	5	0	12	7
[1]	1	15	13	8	10	3	7	4	12	5	6	11	0	14	9	2
[2]	7	11	4	1	9	12	14	2	0	6	10	13	15	3	5	8
[3]	2	1	14	7	4	10	8	13	15	12	9	0	3	5	6	11

```
When 1 =>
        Case column is
            ...
        End case
    ...
End case;
```

Obviously this is quite cumbersome, but also very easy to code automatically using a simple code generator and offers the possibility of the synthesis tool carrying out logic optimization and providing a much more efficient implementation than a memory block.

Validation of DES

In order to validate the implementation of DES, a set of test vectors can be used (i.e. plaintext/ciphertext pairs) to ensure that the correct processing is taking place. A suitable set of test vectors is given below:

Plaintext =   4E6F772069732074 68652074696D6520
              666F7220616C6C20

Ciphertext =   3FA40E8A984D4815 6A271787AB8883F9
               893D51EC4B563B53

In this case the key to be used is 0123456789ABCDEF

Each of the groups of hexadecimal characters is represented by 7-bit ASCII and adding an extra bit.

Advanced Encryption Standard

In the 1997, the US NIST published a request for information regarding the creation of a new AES for non-classified government documents. The call also stipulated that the AES would specify an unclassified, publicly disclosed encryption algorithm(s), available royalty free, worldwide. In addition, the algorithm(s) must implement symmetric key cryptography as a block cipher and (at a minimum) support block sizes of 128 bits and key sizes of 128, 192, and 256 bits.

After an open competition, the Rijndael algorithm was chosen as the winner and implemented as the AES standard. Rijndael allows key and block sizes to be 128, 192 or 256 bits. AES allows the same key sizes, but operates using a block size of 128 bits. The algorithm operates in a similar way to DES, with 10 rounds of confusion and diffusion operators (shuffling and mixing) block at a time. Each

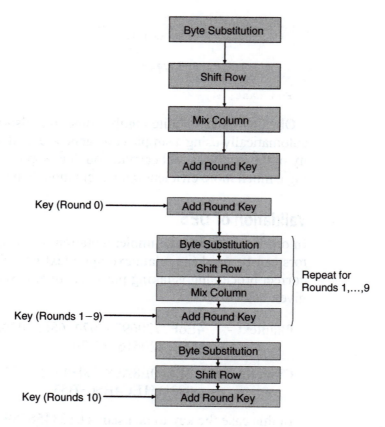

Figure 32
AES Round
Structure

Figure 33
AES Structure

round has a separate key, generated from the overall key. The round structure is shown in Figure 32.

The overall AES structure is given in Figure 33.

Each block consists of 128 bits, and these are divided into 16, 8 bit bytes. Each of the operations acts upon these 8 bit bytes in a 4×4 matrix:

$$\begin{pmatrix} a_{0,0} & a_{0,1} & a_{0,2} & a_{0,3} \\ a_{1,0} & a_{1,1} & a_{1,2} & a_{1,3} \\ a_{2,0} & a_{2,1} & a_{2,2} & a_{2,3} \\ a_{3,0} & a_{3,1} & a_{3,2} & a_{3,3} \end{pmatrix} \tag{1}$$

Note that each (a_i,j) is an 8-bit byte, viewed as elements of $GF(2^8)$. The arithmetic operators take advantage of the Galois Field rules defined in the Rijndael algorithm, an example is that of addition that is implemented by XOR.

Multiplication is more complicated, but each byte has the multiplicative inverse such that $b.b' = 00000001$ (apart from 00000000, whose multiplicative inverse is 00000000).

The model of the finite field $GF(2^8)$ depends on the choice of an irreducible polynomial of degree 8, which for Rijndael is:

$$X^8 + X^4 + X^3 + 1 \qquad (2)$$

Each of the round operations requires a specific mathematical exploration. Taking each in turn:

Byte Substitution requires that for each input data block $a(3,3)$, we look up a table of substitutions and replace the bytes to produce a new matrix $b(3,3)$. The way it works, is that for each input byte **abcdefgh**, we look up row **abcd** and column **efgh** and use the byte at that location in the output.

$$\begin{pmatrix} a_{0,0} & a_{0,1} & a_{0,2} & a_{0,3} \\ a_{1,0} & a_{1,1} & a_{1,2} & a_{1,3} \\ a_{2,0} & a_{2,1} & a_{2,2} & a_{2,3} \\ a_{3,0} & a_{3,1} & a_{3,2} & a_{3,3} \end{pmatrix} \Rightarrow Byte\ substitution$$

$$\Rightarrow \begin{pmatrix} b_{0,0} & b_{0,1} & b_{0,2} & b_{0,3} \\ b_{1,0} & b_{1,1} & b_{1,2} & b_{1,3} \\ b_{2,0} & b_{2,1} & b_{2,2} & b_{2,3} \\ b_{3,0} & b_{3,1} & b_{3,2} & b_{3,3} \end{pmatrix} \qquad (3)$$

The complete byte substitution table is defined using the following figure:

```
099 124 119 123 242 107 111 197 048 001 103 043 254 215 171 118
202 130 201 125 250 089 071 240 173 212 162 175 156 164 114 192
183 253 147 038 054 063 247 204 052 165 229 241 113 216 049 021
004 199 035 195 024 150 005 154 007 018 128 226 235 039 178 117
009 131 044 026 027 110 090 160 082 059 214 179 041 227 047 132
083 209 000 237 032 252 177 091 106 203 190 057 074 076 088 207
208 239 170 251 067 077 051 133 069 249 002 127 080 060 159 168
081 163 064 143 146 157 056 245 188 182 218 033 016 255 243 210
205 012 019 236 095 151 068 023 196 167 126 061 100 093 025 115
096 129 079 220 034 042 144 136 070 238 184 020 222 094 011 219
224 050 058 010 073 006 036 092 194 211 172 098 145 149 228 121
231 200 055 109 141 213 078 169 108 086 244 234 101 122 174 008
186 120 037 046 028 166 180 198 232 221 116 031 075 189 139 138
112 062 181 102 072 003 246 014 097 053 087 185 134 193 029 158
225 248 152 017 105 217 142 148 155 030 135 233 206 085 040 223
140 161 137 013 191 230 066 104 065 153 045 015 176 084 187 022
```

AES Byte Substitution Table

For Example: If the input data byte was 7A, then this in binary terms is

$$0111\ 1010$$

So the row required is 7 (0111) and the column required is A (1010). From this we can read off the resulting number from the table:

$$218 = 1101\ 1010 = DA$$

This is illustrated in the figure below:

	0	1	2	3	4	5	6	7	8	9	A	B	C	D	E	F
0	099	124	119	123	242	107	111	197	048	001	103	043	254	215	171	118
1	202	130	201	125	250	089	071	240	173	212	162	175	156	164	114	192
2	183	253	147	038	054	063	247	204	052	165	229	241	113	216	049	021
3	004	199	035	195	024	150	005	154	007	018	128	226	235	039	178	117
4	009	131	044	026	027	110	090	160	082	059	214	179	041	227	047	132
5	083	209	000	237	032	252	177	091	106	203	190	057	074	076	088	207
6	208	239	170	251	067	077	051	133	069	249	002	127	080	060	159	168
7	081	163	064	143	146	157	056	245	188	182	218	033	016	255	243	210
8	205	012	019	236	095	151	068	023	196	167	126	061	100	093	025	115
9	096	129	079	220	034	042	144	136	070	238	184	020	222	094	011	219
A	224	050	058	010	073	006	036	092	194	211	172	098	145	149	228	121
B	231	200	055	109	141	213	078	169	108	086	244	234	101	122	174	008
C	186	120	037	046	028	166	180	198	232	221	116	031	075	189	139	138
D	112	062	181	102	072	003	246	014	097	053	087	185	134	193	029	158
E	225	248	152	017	105	217	142	148	155	030	135	233	206	085	040	223
F	140	161	137	013	191	230	066	104	065	153	045	015	176	084	187	022

We can see that this is a bit shuffling operation that is simply moving bytes around in a publicly defined manner that does not have anything to do with a key.

Also note that the individual bits within the byte are not changed *per se*. This is a bytewise operation.

The Shift Row function is essentially a set of cyclic shifts to the left with offsets of 0,1,2,3 respectively.

$$\begin{pmatrix} c_{0,0} & c_{0,1} & c_{0,2} & c_{0,3} \\ c_{1,0} & c_{1,1} & c_{1,2} & c_{1,3} \\ c_{2,0} & c_{2,1} & c_{2,2} & c_{2,3} \\ c_{3,0} & c_{3,1} & c_{3,2} & c_{3,3} \end{pmatrix} = \begin{pmatrix} b_{0,0} & b_{0,1} & b_{0,2} & b_{0,3} \\ b_{1,1} & b_{1,2} & b_{1,3} & b_{1,0} \\ b_{2,2} & b_{2,3} & b_{2,0} & b_{2,1} \\ b_{3,3} & b_{3,0} & b_{3,1} & b_{3,2} \end{pmatrix} \qquad (4)$$

The Mix Columns function is a series of specific multiplications:

$$\begin{pmatrix} d_{0,0} & d_{0,1} & d_{0,2} & d_{0,3} \\ d_{1,0} & d_{1,1} & d_{1,2} & d_{1,3} \\ d_{2,0} & d_{2,1} & d_{2,2} & d_{2,3} \\ d_{3,0} & d_{3,1} & d_{3,2} & d_{3,3} \end{pmatrix} = \begin{pmatrix} '02' & '03' & '01' & '01' \\ '01' & '02' & '03' & '01' \\ '01' & '01' & '02' & '03' \\ '03' & '01' & '01' & '02' \end{pmatrix}$$

$$ * \begin{pmatrix} c_{0,0} & c_{0,1} & c_{0,2} & c_{0,3} \\ c_{1,0} & c_{1,1} & c_{1,2} & c_{1,3} \\ c_{2,0} & c_{2,1} & c_{2,2} & c_{2,3} \\ c_{3,0} & c_{3,1} & c_{3,2} & c_{3,3} \end{pmatrix} \quad (5)$$

Where:

'01' = 00000001

'02' = 00000010

'03' = 00000011

All multiplications are $GF(2^8)$ and this transformation is invertible.

The final operation in each round is to add the key using the following function:

$$\begin{pmatrix} e_{0,0} & e_{0,1} & e_{0,2} & e_{0,3} \\ e_{1,0} & e_{1,1} & e_{1,2} & e_{1,3} \\ e_{2,0} & e_{2,1} & e_{2,2} & e_{2,3} \\ e_{3,0} & e_{3,1} & e_{3,2} & e_{3,3} \end{pmatrix} = \begin{pmatrix} d_{0,0} & d_{0,1} & d_{0,2} & d_{0,3} \\ d_{1,0} & d_{1,1} & d_{1,2} & d_{1,3} \\ d_{2,0} & d_{2,1} & d_{2,2} & d_{2,3} \\ d_{3,0} & d_{3,1} & d_{3,2} & d_{3,3} \end{pmatrix}$$

$$\oplus \begin{pmatrix} k_{0,0} & k_{0,1} & k_{0,2} & k_{0,3} \\ k_{1,0} & k_{1,1} & k_{1,2} & k_{1,3} \\ k_{2,0} & k_{2,1} & k_{2,2} & k_{2,3} \\ k_{3,0} & k_{3,1} & k_{3,2} & k_{3,3} \end{pmatrix} \quad (6)$$

The round keys are generated using the following method. The original key of 128 bits is represented as a 4×4 matrix of bytes (of 8 bits). This can be thought of as 4 columns $W(0)$, $W(1)$, $W(2)$, $W(3)$. Adjoin 40 columns $W(4), \ldots, W(43)$.

Round key for round i consists of columns ($W(i)$, $W(i + 1)$, $W(i + 2)$, $W(i + 3)$). If i is a multiple of 4, $W(i) = W(i - 4) \oplus T(W(i - 1))$ where T is a transformation of a, b, c, d in column $W(i - 1)$:

- Shift cyclically to get b, c, d, a.

- Replace each byte with S box entry using ByteSub, to get e, f, g, h.

- Compute round constant $r(i) = 00000010 \; (i - 4)/4$ in $GF(2^8)$.

- $T(W(i - 1)) = (e \oplus r(i), f, g, h)$

If i is not a multiple of 4,

- $W(i) = W(i - 4) \oplus W(i - 1)$

Implementing AES in VHDL

We have two options for implementing block cipher operations in VHDL. We can use the structural approach (shown in the DES example previously in this chapter), or sometimes it makes sense to define a library of functions and use those to make much simpler models.

In the AES example, we can define a top level entity and architecture that has the bare minimum of structure and is completely defined using functions. This can be especially useful when working with behavioral synthesis software as this allows complete flexibility for architectural optimization:

```
library ieee;
use ieee.std_logic_1164.all;
entity AES is
     port(
          plaintext : in std_logic_vector(127 downto 0);
          keytext : in std_logic_vector(127 downto 0);
          encrypt : in std_logic;
          go : in std_logic;
          ciphertext : out std_logic_vector
            (127 downto 0);
          done : out std_logic := '0'
     );
end;


use work.aes_functions.all;
architecture behaviour of AES is
begin
     process
     begin
          wait until go = '1';
          done <= '0';
          ciphertext <= aes_core(plaintext, keytext,
            encrypt);
          done <= '1';
     end process;
end;
```

In this example, we have the plaintext and keytext inputs defined as 128 bit wide vectors and the ciphertext output is also defined as 128 bits wide. The 'go' flag initiates the encryption and the 'done' flag shows when this has been completed.

Notice that we have a work library defined called aes_functions which encapsulates all the relevant functions for the AES algorithm. The set of functions are defined in a package (aes_functions) and this VHDL is given below:

```vhdl
library ieee;
use ieee.std_logic_1164.all;
use ieee.numeric_std.all;
package aes_functions is

        constant nr : integer := 10;
        constant nb : integer := 4;
        constant nk : integer := 4;

        subtype vec1408 is std_logic_vector(1407 downto 0);
        subtype vec128 is std_logic_vector(127 downto 0);
        subtype vec64 is std_logic_vector(63 downto 0);
        subtype vec32 is std_logic_vector(31 downto 0);
        subtype vec16 is std_logic_vector(15 downto 0);
        subtype vec8 is std_logic_vector(7 downto 0);

    subtype int9 is integer range 0 to 9;

        function input_output (input : vec128)   return
          vec128;
        function sBox (pt : vec8)     return vec8;
        function subBytes (plaintext : vec128)   return
          vec128;
        function shiftRows (plaintext : vec128)   return
          vec128;
        function ffmul(pt : vec8; mul : vec8)   return vec8;
        function mixCL(  l0 : vec8; l1 : vec8; l2 : vec8;
          l3 : vec8)   return vec8;
        function mixColumns(pt : vec128)   return vec128;
        function rcon (input : int9)   return vec8;
        function aes_keyexpansion(key : vec128)   return
          vec1408;
        function aes_core (signal plaintext : vec128;
          signal keytext : vec128; signal encrypt :
          std_logic) return vec128;

    end;

    library ieee;
    use ieee.std_logic_1164.all;
    use ieee.numeric_std.all;
    package body aes_functions is
```

```
------------------------------------------------------
----------------------------
    function subBytes (plaintext : vec128)
-- moods inline
    return vec128 is
        variable ciphertext : vec128;
    begin
    ciphertext := sBox(plaintext(127 DOWNTO 120)) &
                  sBox(plaintext(119 DOWNTO 112)) &
                  sBox(plaintext(111 DOWNTO 104)) &
                  sBox(plaintext(103 DOWNTO 96)) &
                  sBox(plaintext(95 DOWNTO 88)) &
                  sBox(plaintext(87 DOWNTO 80)) &
                  sBox(plaintext(79 DOWNTO 72)) &
                  sBox(plaintext(71 DOWNTO 64)) &
                  sBox(plaintext(63 DOWNTO 56)) &
                  sBox(plaintext(55 DOWNTO 48)) &
                  sBox(plaintext(47 DOWNTO 40)) &
                  sBox(plaintext(39 DOWNTO 32)) &
                  sBox(plaintext(31 DOWNTO 24)) &
                  sBox(plaintext(23 DOWNTO 16)) &
                  sBox(plaintext(15 DOWNTO 8)) &
                  sBox(plaintext(7 DOWNTO 0));
    return ciphertext;
    end;


------------------------------------------------------
----------------------------
    function shiftRows (plaintext : vec128)
-- moods inline
    return vec128 is
        variable ciphertext : vec128;
    begin
        --line 0 (the first): no shift
    ciphertext := plaintext(31 DOWNTO 24) &
                  plaintext(55 DOWNTO 48) &
                  plaintext(79 DOWNTO 72) &
                  plaintext(103 DOWNTO 96) &
                  plaintext(127 DOWNTO 120) &
                  plaintext(23 DOWNTO 16) &
                  plaintext(47 DOWNTO 40) &
                  plaintext(71 DOWNTO 64) &
                  plaintext(95 DOWNTO 88) &
                  plaintext(119 DOWNTO 112) &
                  plaintext(15 DOWNTO 8) &
                  plaintext(39 DOWNTO 32) &
                  plaintext(63 DOWNTO 56) &
                  plaintext(87 DOWNTO 80) &
                  plaintext(111 DOWNTO 104) &
                  plaintext(7 DOWNTO 0);
    return ciphertext;
    end;
```

```
------------------------------------------------------
---------------------------
    function tableLog (input : vec8)
-- moods inline
    return vec8 is
        variable output : vec8;
        type table256 is array(0 to 255) of vec8;
        constant pt_256 : table256 := (
        -- moods rom
            X"00",   X"00",   X"19",   X"01",   X"32",
            X"02",   X"1a",   X"c6",   X"4b",
            X"c7",   X"1b",   X"68",   X"33",
            X"ee",   X"df",   X"03",   X"64",

            X"04",   X"e0",   X"0e",   X"34",
            X"8d",   X"81",   X"ef",   X"4c",
            X"71",   X"08",   X"c8",   X"f8",
            X"69",   X"1c",   X"c1",   X"7d",

            X"c2",   X"1d",   X"b5",   X"f9",
            X"b9",   X"27",   X"6a",   X"4d",
            X"e4",   X"a6",   X"72",   X"9a",
            X"c9",   X"09",   X"78",   X"65",

            X"2f",   X"8a",   X"05",   X"21",
            X"0f",   X"e1",   X"24",   X"12",
            X"f0",   X"82",   X"45",   X"35",
            X"93",   X"da",   X"8e",   X"96",

            X"8f",   X"db",   X"bd",   X"36",
            X"d0",   X"ce",   X"94",   X"13",
            X"5c",   X"d2",   X"f1",   X"40",
            X"46",   X"83",   X"38",   X"66",
            X"dd",   X"fd",   X"30",   X"bf",
            X"06",   X"8b",   X"62",   X"b3",
            X"25",   X"e2",   X"98",   X"22",
            X"88",   X"91",   X"10",   X"7e",

            X"6e",   X"48",   X"c3",   X"a3",
            X"b6",   X"1e",   X"42",   X"3a",
            X"6b",   X"28",   X"54",   X"fa",
            X"85",   X"3d",   X"ba",   X"2b",

            X"79",   X"0a",   X"15",   X"9b",
            X"9f",   X"5e",   X"ca",   X"4e",
            X"d4",   X"ac",   X"e5",   X"f3",
            X"73",   X"a7",   X"57",   X"af",

            X"58",   X"a8",   X"50",   X"f4",
            X"ea",   X"d6",   X"74",   X"4f",
            X"ae",   X"e9",   X"d5",   X"e7",
            X"e6",   X"ad",   X"e8",   X"2c",
```

```
                      X"d7",   X"75",   X"7a",   X"eb",
                      X"16",   X"0b",   X"f5",   X"59",
                      X"cb",   X"5f",   X"b0",   X"9c",
                      X"a9",   X"51",   X"a0",   X"7f",

                      X"0c",   X"f6",   X"6f",   X"17",
                      X"c4",   X"49",   X"ec",   X"d8",
                      X"43",   X"1f",   X"2d",   X"a4",
                      X"76",   X"7b",   X"b7",   X"cc",

                      X"bb",   X"3e",   X"5a",   X"fb",
                      X"60",   X"b1",   X"86",   X"3b",
                      X"52",   X"a1",   X"6c",   X"aa",
                      X"55",   X"29",   X"9d",   X"97",

                      X"b2",   X"87",   X"90",   X"61",
                      X"be",   X"dc",   X"fc",   X"bc",
                      X"95",   X"cf",   X"cd",   X"37",
                      X"3f",   X"5b",   X"d1",   X"53",

                      X"39",   X"84",   X"3c",   X"41",
                      X"a2",   X"6d",   X"47",   X"14",
                      X"2a",   X"9e",   X"5d",   X"56",
                      X"f2",   X"d3",   X"ab",   X"44",

                      X"11",   X"92",   X"d9",   X"23",
                      X"20",   X"2e",   X"89",   X"b4",
                      X"7c",   X"b8",   X"26",   X"77",
                      X"99",   X"e3",   X"a5",   X"67",

                      X"4a",   X"ed",   X"de",   X"c5",
                      X"31",   X"fe",   X"18",   X"0d",
                      X"63",   X"8c",   X"80",   X"c0",
                      X"f7",   X"70",   X"07" );
begin
        output := pt_256(TO_INTEGER(UNSIGNED(input)));
        return output;
end;


---------------------------------------------------------
----------------------------
    function tableExp (input : vec8)
-- moods inline
    return vec8 is
        variable output : vec8;
        type table256 is array(0 to 255) of vec8;
        constant pt_256 : table256 := (
        -- moods rom
                X"01",   X"03",   X"05",   X"0f",
                X"11",   X"33",   X"55",   X"ff",
                X"1a",   X"2e",   X"72",   X"96",
                X"a1",   X"f8",   X"13",   X"35",
```

```
X"5f",   X"e1",   X"38",   X"48",
X"d8",   X"73",   X"95",   X"a4",
X"f7",   X"02",   X"06",   X"0a",
X"1e",   X"22",   X"66",   X"aa",


X"e5",   X"34",   X"5c",   X"e4",
X"37",   X"59",   X"eb",   X"26",
X"6a",   X"be",   X"d9",   X"70",
X"90",   X"ab",   X"e6",   X"31",


X"53",   X"f5",   X"04",   X"0c",
X"14",   X"3c",   X"44",   X"cc",
X"4f",   X"d1",   X"68",   X"b8",
X"d3",   X"6e",   X"b2",   X"cd",


X"4c",   X"d4",   X"67",   X"a9",
X"e0",   X"3b",   X"4d",   X"d7",
X"62",   X"a6",   X"f1",   X"08",
X"18",   X"28",   X"78",   X"88",


X"83",   X"9e",   X"b9",   X"d0",
X"6b",   X"bd",   X"dc",   X"7f",
X"81",   X"98",   X"b3",   X"ce",
X"49",   X"db",   X"76",   X"9a",


X"b5",   X"c4",   X"57",   X"f9",
X"10",   X"30",   X"50",   X"f0",
X"0b",   X"1d",   X"27",   X"69",
X"bb",   X"d6",   X"61",   X"a3",


X"fe",   X"19",   X"2b",   X"7d",
X"87",   X"92",   X"ad",   X"ec",
X"2f",   X"71",   X"93",   X"ae",
X"e9",   X"20",   X"60",   X"a0",


X"fb",   X"16",   X"3a",   X"4e",
X"d2",   X"6d",   X"b7",   X"c2",
X"5d",   X"e7",   X"32",   X"56",
X"fa",   X"15",   X"3f",   X"41",


X"c3",   X"5e",   X"e2",   X"3d",
X"47",   X"c9",   X"40",   X"c0",
X"5b",   X"ed",   X"2c",   X"74",
X"9c",   X"bf",   X"da",   X"75",


X"9f",   X"ba",   X"d5",   X"64",
X"ac",   X"ef",   X"2a",   X"7e",
X"82",   X"9d",   X"bc",   X"df",
X"7a",   X"8e",   X"89",   X"80",
```

```
                              X"9b",   X"b6",   X"c1",   X"58",
                              X"e8",   X"23",   X"65",   X"af",
                              X"ea",   X"25",   X"6f",   X"b1",
                              X"c8",   X"43",   X"c5",   X"54",


                              X"fc",   X"1f",   X"21",   X"63",
                              X"a5",   X"f4",   X"07",   X"09",
                              X"1b",   X"2d",   X"77",   X"99",
                              X"b0",   X"cb",   X"46",   X"ca",


                              X"45",   X"cf",   X"4a",   X"de",
                              X"79",   X"8b",   X"86",   X"91",
                              X"a8",   X"e3",   X"3e",   X"42",
                              X"c6",   X"51",   X"f3",   X"0e",


                              X"12",   X"36",   X"5a",   X"ee",
                              X"29",   X"7b",   X"8d",   X"8c",
                              X"8f",   X"8a",   X"85",   X"94",
                              X"a7",   X"f2",   X"0d",   X"17",


                              X"39",   X"4b",   X"dd",   X"7c",
                              X"84",   X"97",   X"a2",   X"fd",
                              X"1c",   X"24",   X"6c",   X"b4",
                              X"c7",   X"52",   X"f6",   X"01");
        begin
                output := pt_256(TO_INTEGER(UNSIGNED(input)));
        return output;
        end;


        --------------------------------------------------------------
        ----------------------
        function ffmul(pt : vec8; mul : vec8)
        -- moods inline
        return vec8 is
                        -- variable res : vec8;
                        variable tablogpt : vec8;
                        variable tablogmul : vec8;
                        variable tablogpt8 : unsigned(8 downto 0);
                        variable tablogmul8 : unsigned(8 downto 0);
                        variable carrie : std_logic_vector (8 downto 0);
                        variable power : vec8;
        variable result: vec8;
          begin
        tablogpt := tableLog(pt);
        tablogmul := tableLog(mul);


        tablogpt8 := unsigned("0" & tablogpt);
        tablogmul8 := unsigned("0" & tablogmul);


        carrie := std_logic_vector(tablogmul8 + tablogpt8);
```

```
        if pt = X"00" or mul = X"00" then
                result := X"00";
      elsif carrie(8) = '1' or carrie(7 DOWNTO 0) =
        X"ff" then -- mod 255
            power := std_logic_vector(unsigned(carrie
            (7 DOWNTO 0)) + 1); -- power = power - 255
            result := tableExp(power);
      else
            power := carrie(7 DOWNTO 0);
      result := tableExp(power);
      end if;
   return result;
      end;
```

```
---------------------------------------------------------
------------------------
   function mixCL(    l0 : vec8; l1 : vec8; l2 : vec8;
      l3 : vec8)
-- moods inline
   return vec8 is
        variable ct : vec8;
   begin

            ct := ffmul(l0, X"02") xor ffmul(l1, X"01")
              xor ffmul(l2, X"01") xor ffmul(l3, X"03");

            return ct;

   end;
```

```
---------------------------------------------------------
------------------------
   function mixColumns(pt : vec128)
-- moods inline
   return vec128 is
        variable ct : vec128;
   begin

            ct := mixCL(pt(127 DOWNTO 120), pt(119 DOWNTO
              112), pt(111 DOWNTO 104), pt(103 DOWNTO 96)) &
            mixCL(pt(119 DOWNTO 112), pt(111 DOWNTO 104),
              pt(103 DOWNTO 96), pt(127 DOWNTO 120)) &
            mixCL(pt(111 DOWNTO 104), pt(103 DOWNTO 96),
              pt(127 DOWNTO 120), pt(119 DOWNTO 112)) &
            mixCL(pt(103 DOWNTO 96), pt(127 DOWNTO 120),
              pt(119 DOWNTO 112), pt(111 DOWNTO 104)) &

            mixCL(pt(95 DOWNTO 88), pt(87 DOWNTO 80),
               pt(79 DOWNTO 72), pt(71 DOWNTO 64)) &
            mixCL(pt(87 DOWNTO 80), pt(79 DOWNTO 72), pt(71
              DOWNTO 64), pt(95 DOWNTO 88)) &
            mixCL(pt(79 DOWNTO 72), pt(71 DOWNTO 64), pt(95
              DOWNTO 88), pt(87 DOWNTO 80)) &
            mixCL(pt(71 DOWNTO 64), pt(95 DOWNTO 88), pt(87
              DOWNTO 80), pt(79 DOWNTO 72)) &
```

```
                    mixCL(pt(63 DOWNTO 56), pt(55 DOWNTO 48), pt(47
                        DOWNTO 40), pt(39 DOWNTO 32)) &
                    mixCL(pt(55 DOWNTO 48), pt(47 DOWNTO 40), pt(39
                        DOWNTO 32), pt(63 DOWNTO 56)) &
                    mixCL(pt(47 DOWNTO 40), pt(39 DOWNTO 32), pt(63
                        DOWNTO 56), pt(55 DOWNTO 48)) &
                    mixCL(pt(39 DOWNTO 32), pt(63 DOWNTO 56), pt(55
                        DOWNTO 48), pt(47 DOWNTO 40)) &

                   mixCL(pt(31 DOWNTO 24), pt(23 DOWNTO 16), pt(15
                        DOWNTO 8), pt(7 DOWNTO 0)) &
                    mixCL(pt(23 DOWNTO 16), pt(15 DOWNTO 8), pt(7
                        DOWNTO 0), pt(31 DOWNTO 24)) &
                    mixCL(pt(15 DOWNTO 8), pt(7 DOWNTO 0), pt(31
                        DOWNTO 24), pt(23 DOWNTO 16)) &
                    mixCL(pt(7 DOWNTO 0), pt(31 DOWNTO 24), pt(23
                        DOWNTO 16), pt(15 DOWNTO 8));

                return ct;
        end;

    -----------------------------------------------------------
    ---------------------------
    function input_output (input : vec128)
-- moods inline
    return vec128 is
            variable output : vec128;
    function flip(input:vec32) return vec32 is
    -- moods inline
    begin
      return input(7 DOWNTO 0) & input(15 DOWNTO 8) &
        input(23 DOWNTO 16) & input(31 DOWNTO 24);
    end;

    begin
            return flip(input(127 downto 96)) & flip(input
                (95 downto 64)) & flip(input(63 downto 32)) &
                flip(input(31 downto 0));
        end;

    -----------------------------------------------------------
    ---------------------------
    function aes_keyexpansion(key : vec128)
-- moods inline
    return vec1408 is
            variable iok : vec128;
            variable er0,er1,er2,er3,er4,er5,er6,er7,
            er8,er9:vec128;
        -- variable zero: vec128;
        -- variable expandedkeys: vec1408;

    function exp_round(input : vec128; round: int9)
    return vec128 is
-- moods inline
        variable r1,r2,r3,r4,r5: vec32;
```

```
    begin
        r1 := sBox(input(7 downto 0)) &
            sBox(input(31 downto 24)) &
            sBox(input(23 downto 16)) &
            (sBox(input(15 downto 8)) xor rcon(round));

        r2 := input(127 downto 96) xor r1;
        r3 := input(95 downto 64) xor r2;
        r4 := input(63 downto 32) xor r3;
        r5 := input(31 downto 0) xor r4;
        return r2 & r3 & r4 & r5;
    end;

    begin
      -- First Round
      iok := input_output(key);
      -- other rounds
      er9 := exp_round(iok,9);
      er8 := exp_round(er9,8);
      er7 := exp_round(er8,7);
      er6 := exp_round(er7,6);
      er5 := exp_round(er6,5);
      er4 := exp_round(er5,4);
      er3 := exp_round(er4,3);
      er2 := exp_round(er3,2);
      er1 := exp_round(er2,1);
      er0 := exp_round(er1,0);

      return (iok & er9 & er8 & er7 & er6 & er5 & er4 &
        er3 & er2 & er1 & er0);
    end;
-----------------------------------------------------------
---------------------------
    function aes_core (signal plaintext : vec128; signal
    keytext : vec128; signal encrypt : std_logic)
-- moods inline
    return vec128 is
        variable rk0 : vec128;
        variable ciphertext, expkey : vec128;
        variable ct1, ct2,ct3,ct4,ct5,ct6,ct7,ct8: vec128;
        variable expandedkeys : vec1408;
    begin
        -- expanded key schedule
        expandedkeys := aes_keyexpansion(keytext);

        -- Round 0
        ct1 := input_output(plaintext) xor
          expandedkeys(1407 downto 1280);

    -- Round 1 to Nr-1
    -- for i in 1 to Nr-1 loop
```

```
        for i in 1 to 9 loop
                    ct2 := subBytes(ct1);
                    ct3 := shiftRows(ct2);
                    ct4 := mixColumns(ct3);
        case(i) is
          when 1 => expkey := expandedkeys(1279 downto 1152);
          when 2 => expkey := expandedkeys(1151 downto 1024);
          when 3 => expkey := expandedkeys(1023 downto 896);
          when 4 => expkey := expandedkeys(895 downto 768);
          when 5 => expkey := expandedkeys(767 downto 640);
          when 6 => expkey := expandedkeys(639 downto 512);
          when 7 => expkey := expandedkeys(511 downto 384);
          when 8 => expkey := expandedkeys(383 downto 256);
          when 9 => expkey := expandedkeys(255 downto 128);
          when others => null;
        end case;
                      ct1 := ct4 xor expkey;
                end loop;

                -- Final Round Nr=10
                ct5 := subBytes(ct1);
                ct6 := shiftRows(ct5);
                ct7 := ct6 xor expandedkeys(1407-128*Nr downto
                  1280-128*Nr);

                ciphertext := input_output(ct7);
                return ciphertext;
        end;

    -----------------------------------------------------------
    ---------------------------
    function rcon (input : int9)
-- moods inline
    return vec8 is
    type rcont_t is array(0 to 9) of vec8;
    constant table_rcon: rcont_t := (
    -- moods rom
    X"36", X"1b", X"80", X"40", X"20", X"10", X"08", X"04",
      X"02", X"01");
    begin
            return table_rcon(input);
    end;

    -----------------------------------------------------------
    ---------------------------
    function sBox (pt : vec8)
-- moods inline
    return vec8 is
        variable ct : vec8;
        type table256 is array(0 to 255) of vec8;
        constant pt_256 : table256 := (
        -- moods rom
    X"63", X"7c", X"77", X"7b", X"f2", X"6b", X"6f", X"c5",
    X"30", X"01", X"67", X"2b", X"fe", X"d7", X"ab", X"76",
```

```
X"ca", X"82", X"c9", X"7d", X"fa", X"59", X"47", X"f0",
X"ad", X"d4", X"a2", X"af", X"9c", X"a4", X"72", X"c0",
X"b7", X"fd", X"93", X"26", X"36", X"3f", X"f7", X"cc",
X"34", X"a5", X"e5", X"f1", X"71", X"d8", X"31", X"15",
X"04", X"c7", X"23", X"c3", X"18", X"96", X"05", X"9a",
X"07", X"12", X"80", X"e2", X"eb", X"27", X"b2", X"75",
X"09", X"83", X"2c", X"1a", X"1b", X"6e", X"5a", X"a0",
X"52", X"3b", X"d6", X"b3", X"29", X"e3", X"2f", X"84",
X"53", X"d1", X"00", X"ed", X"20", X"fc", X"b1", X"5b",
X"6a", X"cb", X"be", X"39", X"4a", X"4c", X"58", X"cf",
X"d0", X"ef", X"aa", X"fb", X"43", X"4d", X"33", X"85",
X"45", X"f9", X"02", X"7f", X"50", X"3c", X"9f", X"a8",
X"51", X"a3", X"40", X"8f", X"92", X"9d", X"38", X"f5",
X"bc", X"b6", X"da", X"21", X"10", X"ff", X"f3", X"d2",
X"cd", X"0c", X"13", X"ec", X"5f", X"97", X"44", X"17",
X"c4", X"a7", X"7e", X"3d", X"64", X"5d", X"19", X"73",
X"60", X"81", X"4f", X"dc", X"22", X"2a", X"90", X"88",
X"46", X"ee", X"b8", X"14", X"de", X"5e", X"0b", X"db",
X"e0", X"32", X"3a", X"0a", X"49", X"06", X"24", X"5c",
X"c2", X"d3", X"ac", X"62", X"91", X"95", X"e4", X"79",
X"e7", X"c8", X"37", X"6d", X"8d", X"d5", X"4e", X"a9",
X"6c", X"56", X"f4", X"ea", X"65", X"7a", X"ae", X"08",
X"ba", X"78", X"25", X"2e", X"1c", X"a6", X"b4", X"c6",
X"e8", X"dd", X"74", X"1f", X"4b", X"bd", X"8b", X"8a",
X"70", X"3e", X"b5", X"66", X"48", X"03", X"f6", X"0e",
X"61", X"35", X"57", X"b9", X"86", X"c1", X"1d", X"9e",
X"e1", X"f8", X"98", X"11", X"69", X"d9", X"8e", X"94",
X"9b", X"1e", X"87", X"e9", X"ce", X"55", X"28", X"df",
X"8c", X"a1", X"89", X"0d", X"bf", X"e6", X"42", X"68",
X"41", X"99", X"2d", X"0f", X"b0", X"54", X"bb", X"16");
begin
    ct := pt_256(TO_INTEGER(UNSIGNED(pt)));
    return ct;
end;

------------------------------------------------------
end;
```

After the functions and top level entity have been defined, we can implement a test bench that applies a set of test data to the inputs and verifies that the correct output has been obtained. Notice that we use the assertion technique to identify correct operation:

```
library ieee;
use ieee.std_logic_1164.all;
entity testAES is
end;

library ieee;
use ieee.std_logic_1164.all;
```

```vhdl
use work.aes_functions.all;
architecture behaviour of testAES is

    component aes
            port(
                    plaintext : in std_logic_vector (127
                      downto 0);
                    keytext : in std_logic_vector(127
                      downto 0);
                    encrypt : in std_logic;
                    go : in std_logic;
                    ciphertext : out std_logic_vector
                      (127 downto 0);
                    done : out std_logic
              );
    end component;

    for all : aes use entity work.aes;

    signal plaintext :   std_logic_vector(127 downto 0);
    signal keytext :     std_logic_vector(127 downto 0);
    signal encrypt : std_logic;
    signal go : std_logic := '0';
    signal ciphertext : std_logic_vector(127 downto 0);
    signal done : std_logic;
    signal ok : std_logic := '0';
begin
    plaintext <=   X"00000000000000000000000000000000",
      X"3243f6a8885a308d313198a2e0370734" after 50 ns ;
    keytext <=   X"00000000000000000000000000000000",
      X"2b7e151628aed2a6abf7158809cf4f3c" after 100 ns;

    process (ciphertext)
                variable ct : std_logic_vector(127
                  downto 0) :=
                  X"3925841d02dc09fbdc118597196a0b32";
    begin
          assert ct = ciphertext
                    report "Test vectors do not match"
                    severity note;
                 assert not (ct = ciphertext)
                        report "Test vectors Matched"
                        severity note;
    end process;

    process
    begin
                go <= not go after 20 ns;
    end process;

    DUT : aes port map (plaintext, keytext, encrypt,
      go, ciphertext, done);
end;
```

Summary

This chapter shows how two standard block ciphers can be implemented in VHDL. Both of these algorithms are in common usage today and in operational hardware. There are numerous other methods, as security requires a constant evolution of encryption techniques and no doubt more robust and secure methods will emerge that require implementation in VHDL.

10
Memory

Introduction

If we consider SDRAM (Synchronous Dynamic Random Access Memory), the key aspects of this type of memory to consider are:

1. This type of DRAM (Dynamic RAM) relies on transistor capacitance on gates to store data.

2. DRAM is much more compact than SRAM (Static RAM).

3. DRAM cannot be synthesized – you need a separate DRAM chip.

4. SDRAM requires a synchronization clock that is consistent with the rest of the hardware system (it is designed to operate with microprocessors).

5. DRAM data must be refreshed as it is stored charge and decays after a certain time.

6. DRAM is slower than SRAM.

Static RAM (SRAM) can be considered in a similar way to a Read Only Memory (ROM) chip and it also has (differing) key aspects of behavior to consider:

1. Memory cells are based on standard latches.

2. SRAM is fast.

3. SRAM is less compact than DRAM (or SDRAM).

4. SRAM can be synthesized on an Field Programmable Gate Array (FPGA) – so is ideal for small, fast registers or memory blocks.

Statics RAM is essentially asynchronous, but can be modified to behave synchronously (as SDRAM is the synchronous equivalent of DRAM), and this is often called Synchronous RAM.

Flash Memory is useful to consider at this point, even though its operation is fundamentally different from the memory types considered thus far, simply because it is easy to use and is commonly available on many FPGA development boards.

Flash Memory is essentially a form of EEPROM – electrically programmable ROM – that can be used as a form of persistent RAM. Why persistent? In Flash Memory, the device memory is retained even when the power is removed, so it is often used as a form of ROM, which makes it an interesting memory to use on FPGA systems as it could be used to store the FPGA program, but also used as a RAM storage (dynamically) for current data.

Modeling memory in VHDL

Great care must be exercised when modeling memory in VHDL. As some memory cannot be synthesized, if a model is used, it must reflect the correct physical behavior of the real device if it is off chip. This particularly applies to access times and timing violation conditions. If the timing is violated, then the data may be at best suspect and at worst totally useless. The designer can find themselves in the invidious position of having a simulation model that works perfectly, and real hardware that is completely non-functional.

In this chapter, we have used VHDL that does not have any physical delays in any of the models, and these must be added if the models are to be used in a realistic system.

Read Only Memory

ROM is essentially a set of predefined data values in a storage register. The memory has two definitions, first the number of storage areas and second the number of bits. For example, if the memory has 16 storage areas and 8 bits each, the memory is defined as a 16×8 ROM. The basic ROM has one input, the definition of the address to be accessed, and one output, which is a logic vector which is where the data will be put. Consider the entity for a simple behavioral ROM model in VHDL:

```
ENTITY ROM16x8 IS
   PORT (address : IN INTEGER RANGE 0 TO 15;
         dout : OUT std_logic_vector (7 DOWNTO 0));
END ENTITY ROM16x8;
```

As can be seen, the address has been defined as an integer, but the range has been restricted to the range of the ROM. The architecture of the ROM is defined as a fixed array of elements that can be accessed directly. Therefore an example ROM with a set of example data elements could be defined as follows:

```
ARCHITECTURE example OF rom16x8 IS
    TYPE romdata IS ARRAY (0 TO 15)
    OF std_logic_vector( 7 DOWNTO 0);
CONSTANT romvals : romdata := ("00000000",
                              "01010011",
                              "01110010",
                              "01101100",
                              "01110101",
                              "11010111",
                              "11011111",
                              "00111110",
                              "11101100",
                              "10000110",
                              "11111001",
                              "00111001",
                              "01010101",
                              "11110111",
                              "10111111",
                              "11101101");
BEGIN
   data <= romvals(address);
END ARCHITECTURE example;
```

If we wish to use this in an example, we first need to declare the ROM in a VHDL testbench and then specify the address using an integer signal. A sample testbench is given below:

```
library ieee;
use ieee.std_logic_1164.all;

entity testrom is
end entity testrom;

architecture test of testrom is
    signal address : integer := 0;
    signal data : std_logic_vector ( 7 downto 0 );
```

```
begin
    rom16x8: entity work.rom16x8 (example)
        port map ( address, data );
end architecture test;
```

Notice that the IEEE library, std_logic_1164, is required for the std_logic_vector type and the value of the data will depend on the address chosen.

Random Access Memory

A DRAM block has a two-dimensional structure of memory that is divided into a grid structure accessed by a row address and column address. Note that this is asynchronous and therefore has no clock. The implication of being asynchronous is that care must be taken with the timing of the memory access to ensure data integrity throughout the transfer process.

The VHDL model has a single address input and two control signals, RADDR and CADDR, are used to specify the Row and Column Address, respectively. There is also a RW signal that is defined as being write when high and read when low. Finally, the data is put onto the DATA signal which is defined as an INOUT (bidirectional) signal. The resulting entity is given in the VHDL below. In this example, the number of rows is 28 and the number of columns also 28. This gives a total data storage with 16 bits of 1 Mbit

```
ENTITY DRAM1MB IS
  PORT (
      address : IN INTEGER RANGE 0 TO 2**8-1;
      RW : std_logic;
      data : OUT std_logic_vector (15 DOWNTO 0));
END ENTITY DRAM1MB;
```

The architecture is shown in the VHDL below:

```
architecture behav of DRAM1MB is
begin
process (RADDR, CADDR, RW) is
      type dram is array (0 to 2**16 - 1) of
                        std_logic_vector(15 downto 0);
    variable radd: INTEGER range 0 to 2**8 - 1;
    variable madd: INTEGER range 0 to 2**16 - 1;
    variable memory: dram;
begin
  data <= (others => 'Z');
  if falling_edge(RADDR) then
    radd := address;
  elsif falling_edge(CADDR) then
```

```
      madd:=radd*2**18 +Address;
      if RADDR = '0' and RW = '0' then
        memory(madd) := data;
      end if;
    elsif CADDR = '0' and RADDR = '0' and RW = '1' then
      data <= memory(madd);
      end if;
  end process;
  end architecture behav;
```

Using this model a simple testbench can be used to read in a data value to an address, then another value to another address and then the original value read back. The test bench to achieve this is given in the VHDL below.

```
library ieee;
use ieee.std_logic_1164.all;

entity testram is
end entity testram;

architecture test of testram is
    signal address : integer range 0 to 2**8-1 := 0;
    signal rw : std_logic;
    signal c : std_logic;
    signal r : std_logic;
    signal data : std_logic_vector ( 15 downto 0 );
begin

    dram: entity work.dram1mb(behav)
          port map ( address, rw, c, r, data );

    address <= 23 after 0 ns, 47 after 30 ns, 23 after
      90 ns;
    rw <=     '0' after 0 ns, '1' after 90 ns;
    c <=      '1' after 0 ns, '0' after 20 ns,
              '1' after 50 ns, '0' after 70 ns,
              '1' after 90 ns, '0' after 100 ns;
    r <=      '1' after 0 ns, '0' after 10 ns,
              '1' after 40 ns, '0' after 60 ns,
              '1' after 80 ns, '0' after 100 ns;
    data <=   X"1234" after 0 ns, X"5678" after 40 ns;

end architecture test;
```

The results of testing this model can be seen in the waveform (Figure 34), which shows the correct behavior of the address, data and control lines.

It is important to note that the RAM model does not model any of the actual delays that would appear in practice and if this is important to the functionality of the design, then it MUST be added to the model.

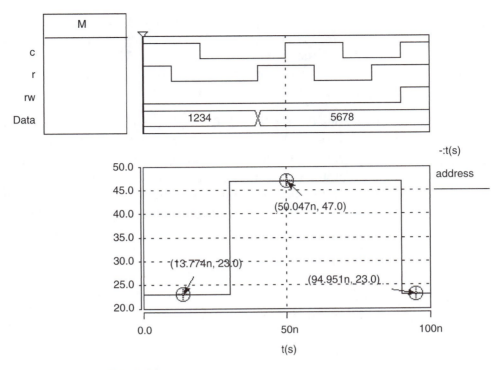

Figure 34
DRAM Simulation Results

Synchronous RAM

In the preceding chapter, we observed how the memory is accessed asynchronously, whereas Synchronous RAM (SRAM) requires a clock. In most practical designs, the RAM will be implemented off chip as a separate memory device, but sometimes it is useful to define a small block of RAM on the FPGA for fast access or local storage close to the hardware device that requires frequent access to a relatively small memory block.

The usual design constraints apply to memory, more so than some other possible functions, as the use of flip-flops to store data without using much of the other logic in a Look-Up Table (LUT) is area intensive. The trade-off, as ever with FPGA design, is whether the potential for improved performance and speed using on board RAM outweighs the increased area required as a result.

From the design perspective, the synchronous RAM VHDL model is very similar to the previously demonstrated basic asynchronous

RAM model. The only difference is that instead of the data being available immediately on the address being applied (or after some short delay), the data in a synchronous RAM is only accessed when the clock edge occurs (rising or falling edge depending on the design required).

If we consider the VHDL for the entity of the SRAM we can see that for a memory size of 2^m and a data bus of 2^n, the following entity is required. The VHDL model has two parameters, m and n. In the default case, the value of m as 10 provides 1024 address words and the number of bits (n) set to 8 gives a total of 8 K bits in the RAM. Obviously this could be made any size, bit this shows the type of calculation required to obtain the specified memory blocks.

```
ENTITY SRAM IS
  GENERIC (
    M : natural := 10;
    N : natural := 8
    );
  PORT (
    clk : in std_logic;
    addr : in std_logic_vector(m-1 downto 0);
    wr : in std_logic;
    d : in std_logic_vector (n-1 downto 0);
    q : out std_logic_vector (n-1 downto 0)
    );
END ENTITY DRAM1MB;
```

Notice that there are two control signals, the clock (clk) and the write enable (wr). We could make the memory synchronous write, synchronous read or a more complex port structure, but in this case, we will show the operation as being synchronous read and write, on the rising edge of the clock. Also, the convention we will use is for the write enable state to be active when wr is low. The resulting VHDL for this synchronous RAM is shown below:

```
Architecture dualport of sram is
    Type sramdata is array (0 to 2**m-1) of
        Std_logic_vector (n-1 downto 0);
    Signal memory : sramdata;
Begin
    Process (clk ) is
    Begin
        If rising_edge(clk) then
            If wr = '0' then
                Memory(to_integer(unsigned(addr))) <= d;
```

```
            Else
                Q <= memory(to_integer(unsigned(addr)));
            End if;
        End if;
    End process;
End architecture dualport;
```

There are several interesting aspects to this model that are worth considering. The first is the access of the memory. If we define the address as a std_logic_vector type in VHDL, then we can't simply use this value to access a specific element of an array. This requires an integer argument. We also cannot simply cast a std_logic_vector type directly to an integer. The first thing we must do is convert the std_logic_vector type to an unsigned number. This is a half way house from std_logic_vector to integer, in that we can use the variable as a number, but it is limited to the same bit resolution as the original std_logic_vector. In this case, clearly this is not an issue as we do not want the address to be larger than the memory otherwise errors will result. The final step is to convert the unsigned type to an integer. This is accomplished using the to_integer function and is the final step to convert the address into the integer form required to access the individual element of the array.

As a consequence of using these numeric functions, we need to also include the ieee standard numeric library in the header of the model as shown:

```
Library ieee;
Use ieee.std_logic_1164.all;
Use ieee.numeric_std.all;
```

It is also worth noting that the read and write functions are mutually exclusive, in that you cannot read from the memory and write to it at the same time. This ensures the integrity of the data. Also note that the read and write functions are both clocked and so the memory is both read and write synchronous.

FLASH memory

As has been discussed previously, FLASH memory is essentially a form of EEPROM (Electrically Erasable and Programmable Read Only Memory). This is slightly different to a standard RAM where the address is given to the memory and depending on the R/W signals, the data is read or written, respectively. A typical set of interface pins for a FLASH memory consists of the following elements:

Pin	Function	Active State
CLE	Command Latch	H, activated on rising_edge(WE)
ALE	Address Latch	H, activated on rising_edge(WE)
CE	Chip Enable	L
RE	Read Enable	Falling_edge(RE)
WE	Write Enable	Rising_edge(WE)
WP	Write Protect	Low
Busy	Ready/Busy	L = busy, H = ready

In addition to these control signals there is of course an address bus and a data bus. To implement this we can use a similar entity to that for a standard RAM block in VHDL:

```
ENTITY FLASH IS
   GENERIC (
     A : natural := 10;
     D : natural := 8
     );
   PORT (
     clk : in std_logic;
     addr : in std_logic_vector(A-1 downto 0);
     data : inout std_logic_vector (D-1 downto 0);
     cle : IN std_logic;
     ale : IN std_logic;
     ce : IN std_logic;
     re : IN std_logic;
     we : IN std_logic;
     wp : IN std_logic;
     busy : OUT std_logic;
     );
END ENTITY FLASH;
```

In most cases we won't need to model the FLASH memory itself, but rather we need to interface to it, so the entity for a FLASH interface controller could be as follows:

```
ENTITY FLASHIF IS
   PORT (
     Clk : IN std_logic;
     read : IN std_logic;
     en : IN std_logic;
     cle : OUT std_logic;
     ale : OUT std_logic;
     ce : OUT std_logic;
     re : OUT std_logic;
```

```
    we : OUT std_logic;
    wp : OUT std_logic;
    busy : IN std_logic;
    );
END ENTITY FLASHIF;
```

A typical architecture for this device could be as follows:

```
Architecture basic of FLASHIF is
Begin
      Process (clk) is
        If busy = '1' then
          If rising_edge(clk) then
              Ce <= en;
              Ale <= '1';
              Cle < '1';
              If read = '0' then
                    We <= '1';
                    Re <= '1';
              Else
                    We <= '0';
                    Re <= '0';
              End if;
              If prog = '0' then
                    Wp <= '0';
              Else
                    Wp <= '1';
              End if;
          End if;
        End if;
    End process;
End architecture basic;
```

This is a basic outline for a flash controller and this will obviously change from device to device.

Summary

This chapter has introduced the important memory types of ROM, asynchronous RAM, FLASH memory and synchronous RAM. It is important to remember that in most cases, large memory blocks will be contained off chip and so it may be necessary to use these models purely for simulation rather than synthesis, but that it is possible to use RAM sparingly on the FPGA itself if absolutely required.

In this case, the trade-off of speed vs. area becomes particularly acute and as such great care must be taken to not make naïve decisions about putting large amounts of memory on the FPGA – as this may take up far too much memory to be practical.

11
PS/2 Mouse Interface

Introduction

The PS/2 mouse is a standard interface to both computers and also many Field Programmable Gate Array (FPGA) development kits. The protocol is a serial one and in this chapter the basics of the protocol will be reviewed and also a simple VHDL interface code to enable the designer to use a mouse, primarily on a standard FPGA development kit.

PS/2 mouse basics

The origins of the PS/2 mouse are back in the 1980s with the proliferation of the IBM Personal Computer (PC). This had the generic name of a 'Personal System' – hence PS and the second version of this was therefore called the PS/2 – and the interface technology has remained under that name ever since.

The PS/2 interface is essentially a custom serial interface with one device supported per connector (unlike the modern USB – Universal Serial Bus – which can handle numerous devices on a single port). The data rate is relatively slow – 40 kbps and the device is powered off a 5 V dc supply.

Unlike the USB approach where devices are generally 'hot swappable' that is they can be plugged in or unplugged at will, the PS/2 device cannot be removed without a system crash or freeze resulting.

The PS/2 mouse supports communication from the mouse to the host and vice versa, and the supply is provided from the host to the mouse in the form of a 5 V line.

PS/2 mouse commands

The PS/2 mouse has a limited set of commands that are essentially either button press commands or mouse movement commands. The standard mouse has a left, middle and right button click command, and the X and Y movement. The X and Y movements are tracked using counters, where the value is relative to the previous value sent by the mouse *not* the absolute position itself.

PS/2 mouse data packets

The PS/2 mouse sends data in serial packets down a data line and this is synchronous with a clock line also on the mouse interface. Each packet consists of 3, 8 bit words where the first word is a configuration word with a set of flags, the second word provides the mouse X movement and the third word provides the mouse Y movement. The description of the mouse bits are given in the table below:

Bit	Byte 1	Byte 2	Byte 3
7	Y overflow	X Movement	Y movement
6	X overflow		
5	Y sign bit		
4	X sign bit		
3	Always 1		
2	Middle Btn		
1	Right Btn		
0	Left Btn		

Each of the movement bytes are defined as 9 bit 2's complement numbers, where the sign bit is defined in byte 1. The range of movement that can be defined is -255 to $+255$ using this approach.

PS/2 operation modes

The PS/2 mouse operates in four basic modes. On power up the mouse goes into a 'reset' mode and this can also be initiated by a reset command from the host which is defined as 0xFF. After reset has been completed, the mouse automatically goes into a 'stream' mode in which the data is streamed back from the mouse to the host. These two modes are he most commonly used modes of operation for most applications, but there are two other used modes which are remote and wrap. These are mostly useful in testing that the interface is operating correctly.

In the reset mode the mouse itself will reset and carry out some basic self checks. The default settings are then defined for the mouse to operate with which are a sample period of 10 ms, a basic resolution of 4 counts per mm, a 1 to 1 scaling and the data reporting option is disabled.

The mouse sends a device ID of 0x00 to the host to let it know that it is not a keyboard or more complex mouse – just a basic PS/2 mouse.

Once the mouse is running it goes into stream mode and the mouse will send packets to the host at the defined sample rate of activity such as mouse movement or button presses. The mouse ONLY sends data when activity is present, otherwise it will do nothing.

If the mouse is asked by the host to go into remote mode, then the mouse only sends data when requested by the host and finally in wrap mode, the mouse sends back every command back to the host (apart from the reset and reset wrap commands).

PS/2 mouse with wheel

A mouse that has a wheel is defined as a separate type of device and so it has a difference device id – 0x03. In this case, after reset, the mouse sends the ID and in the case of a wheel mouse, the data packet is now 4 bytes long and there is an extra byte to provide the wheel movement. This byte only uses the least significant bits in a 2's complement form and therefore has a range of -8 to $+7$.

Basic PS/2 mouse handler VHDL

The simplest form of the VHDL handler could use the mouse clock signal as the system clock and then monitor the data coming from the mouse and this is shown below:

```
Library ieee;
Use ieee.std_logic_1164.all;

Entity psmouse is
        Port (
        Clock : IN std_logic;
        Data : IN std_logic
        );
End entity psmouse;
```

```
Architecture basic of psmouse is
      Signal d : std_logic_vector (23 downto 0);
      Signal byte1 : std_logic_vector (7 downto 0);
      Signal byte2 : std_logic_vector (7 downto 0);
      Signal byte3 : std_logic_vector (7 downto 0);
      Signal index : integer := 23;
Begin
      Process(clock) is
      Begin
            If falling_edge(clock) then
                  D(index)  <= data;
                  If index > 0 then
                        Index <= index-1;
                  Else
                        Byte1 <= d(23 downto 16);
                        Byte2 <= d(15 downto 8);
                        Byte3 <= d(7 downto 0);
                        Index <= 23;
                  End if;
            End if;
      End process;
End architecture basic;
```

This VHDL is very simple and on each falling edge of the clock
the current value of the data is read into the next element of the data
array (d) and when the complete 24 bits packet has been read in (and
index has counted down to zero) then the 3 bytes are then transcribed from the packet.

Modified PS/2 mouse handler VHDL

The trouble with the previous mouse handler is that although syntactically correct, there could be noise on the mouse clock and
data signals leading to an incorrect clocking of the data and so
another approach would be to have a much higher frequency signal clock and to monitor the PS/2 clock as if it was a signal. An
extra check would be to filter the PS/2 clock so that only if there
were a certain number of values the same would the clock be considered to have changed.

```
Library ieee;
Use ieee.std_logic_1164.all;

Entity psmouse is
      Port (
            Clk : IN std_logic;
            Ps2_clock : IN std_logic;
            Data : IN std_logic
      );
End entity psmouse;
```

```
Architecture basic of psmouse is
      Signal clk_internal : std_logic := '0';
      Signal d : std_logic_vector (23 downto 0);
      Signal byte1 : std_logic_vector (7 downto 0);
      Signal byte2 : std_logic_vector (7 downto 0);
      Signal byte3 : std_logic_vector (7 downto 0);
      Signal index : integer : = 23;
Begin
      Process(clock) is
            High : integer : = 0;
            Low : integer : = 0;
      Begin
            If rising_edge(clock) then
            if (ps2_clock = '1') then
                if high = 8 then
                      clk_internal <= '1';
                      high <= 0;
                      low <= 0
                else
                      high <= high + 1;
                end if;
            else
                if low <= 8 then
                      clk_internal <= '0';
                      low <= 0;
                      high <= 0;
                else
                      low <= low + 1;
                end if;
            end if;
            End if;
End process;
Process(clk_internal) is
Begin
      If falling_edge(clk_internal) then
            D(index) <= data;
            If index > 0 then
                  Index <= index - 1;
            Else
                  Byte1 <= d(23 downto 16);
                  Byte2 <= d(15 downto 8);
                  Byte3 <= d(7 downto 0);
                  Index <= 23;
            End if;
      End if;
End process;
End architecture basic;
```

In this case the modified mouse handler waits for 8 consecutive
highs or lows on the clock signal at the higher internal clock rate
of the FPGA and then it will set the internal clock high or low

respectively. Then the same mouse handler routine takes over to manage the data input, this time using the internally generated clock.

Summary

This chapter has shown how to handle a basic PS/2 signal for a mouse and then store the data in 3 bytes for further processing. Two methods are shown for collecting the data, one using the PS/2 clock and the other using a sampled version with a much faster internal clock.

12

PS/2 Keyboard Interface

Introduction

The PS/2 keyboard is a standard interface to both computers and also many Field Programmable Array (FPGA) development kits. The protocol is a serial one and in this chapter the basics of the protocol will be reviewed and also a simple VHDL interface code to enable the designer to use a PS/2 keyboard, primarily on a standard FPGA development kit.

PS/2 keyboard basics

The origins of the PS/2 keyboard are back in the 1980s with the proliferation of the IBM Personal Computer (PC). This had the generic name of a 'Personal System' – hence PS and the second version of this was therefore called the PS/2 – and the interface technology has remained under that name ever since. The keyboard interface evolved from the XT (83 key, 5 pin DIN), through the AT (84–101 key, 5 pin DIN) and eventually settled on the PS/2 (84–101 Key, 6 pin miniDIN).

The PS/2 interface is essentially a custom serial interface with one device supported per connector (unlike the modern USB – Universal Serial Bus – which can handle numerous devices on a single port). The data rate is relatively slow – 40 kbps and the device is powered off a 5 V DC supply.

Unlike the USB approach where devices are generally 'hot swappable' that is they can be plugged in or unplugged at will, the PS/2 device cannot be removed without a system crash or freeze resulting.

The PS/2 keyboard supports communication from the keyboard to the host and vice versa, and the supply is provided from the host to the keyboard in the form of a 5 V line.

Unlike the mouse, the keyboard has an on-board processor that checks the matrix of keys for any key presses and sends the appropriate code down the PS/2 data line.

PS/2 keyboard commands

The keyboard processor has two commands that are sent to the host system when a key is pressed, the 'make' and the 'break' command. Each key has a separate code that is sent in each case. The code that is actually sent to the host has no relationship to the ASCII code of the character sent. It is up to the host code to decode the key command sent. For example, the character '5' has the make code 0x2E and the break code 0xF0,0x2E. Most standard characters have a 1 byte make code and a 2 bytes break code, and extended characters often have 2 bytes make codes and 3 bytes break codes.

If a key is pressed, then the make code is sent periodically until another key is pressed. The rate of this is called the typematic rate and is defined as default at approximately 10 characters per second.

PS/2 keyboard data packets

The PS/2 keyboard sends data in serial packets down a data line and this is synchronous with a clock line also on the mouse interface. Each packet consists of up to 3, 8-bit bytes and this can be decoded by a look-up table for the keyboard scan codes.

PS/2 keyboard operation modes

Basic PS/2 keyboard handler VHDL

The simplest form of the VHDL handler could use the keyboard clock signal as the system clock and then monitor the data coming from the keyboard and this is shown below:

```
Library ieee;
Use ieee.std_logic_1164.all;

Entity pskeyboard is
        Port (
                Clock : IN std_logic;
                Data : IN std_logic
        );
End entity pskeyboard;
```

```
Architecture basic of pskeyboard is
      Signal d : std_logic_vector (23 downto 0);
      Signal byte1 : std_logic_vector (7 downto 0);
      Signal byte2 : std_logic_vector (7 downto 0);
      Signal byte3 : std_logic_vector (7 downto 0);
      Signal index : integer := 23;
Begin
      Process(clock) is
      Begin
            If falling_edge(clock) then
                  D(index) <= data;
                  If index > 0 then
                        Index  <= index-1;
                  Else
                        Byte1 <= d(23 downto 16);
                        Byte2 <= d(15 downto 8);
                        Byte3 <= d(7 downto 0);
                        Index <= 23;
                  End if;
            End if;
      End process;
End architecture basic;
```

This VHDL is very simple and on each falling edge of the clock the current value of the data is read into the next element of the data array (d) and when the complete 24-bit packet has been read in (and index has counted down to zero) then the three bytes are then transcribed from the packet.

Modified PS/2 keyboard handler VHDL

The trouble with the previous keyboard handler is that although syntactically correct, there could be noise on the keyboard clock and data signals leading to an incorrect clocking of the data and so another approach would be to have a much higher frequency signal clock and to monitor the PS/2 clock as if it was a signal. An extra check would be to filter the PS/2 clock so that only if there were a certain number of values the same would the clock be considered to have changed.

```
Library ieee;
Use ieee.std_logic_1164.all;

Entity pskeyboard is
      Port (
            Clk : IN std_logic;
            Ps2_clock : IN std_logic;
            Data : IN std_logic
      );
End entity pskeyboard;
```

```
Architecture basic of pskeyboard is
      Signal clk_internal : std_logic := '0';
      Signal d : std_logic_vector (23 downto 0);
      Signal byte1 : std_logic_vector (7 downto 0);
      Signal byte2 : std_logic_vector (7 downto 0);
      Signal byte3 : std_logic_vector (7 downto 0);
      Signal index : integer := 23;
Begin
      Process(clock) is
            High : integer := 0;
            Low : integer := 0;
      Begin
            If rising_edge(clock) then
            if (ps2_clock = '1') then
                  if high = 8 then
                              clk_internal <= '1';
                              high <= 0;
                              low <= 0
                        else
                              high <= high +1;
                        end if;
                  else
                        if low = 8 then
                              clk_internal <= '0';
                              low <= 0;
                              high <= 0;
                        else
                              low <= low +1;
                        end if;
                  end if;
            End if;
      End process;
Process(clk_internal) is
Begin
      If falling_edge(clk_internal) then
            D(index) <= data;
            If index > 0 then
                  Index <= index-1;
            Else
                  Byte1 <= d(23 downto 16);
                  Byte2 <= d(15 downto 8);
                  Byte3 <= d(7 downto 0);
                  Index <= 23;
            End if;
      End if;
End process;
End architecture basic;
```

In this case the modified keyboard handler waits for 8 consecutive highs or lows on the clock signal at the higher internal clock rate of the FPGA and then it will set the internal clock high or low respectively. Then the same keyboard handler routine takes over to manage the data input, this time using the internally generated clock.

Summary

This chapter has shown how to handle a basic PS/2 signal for a keyboard and then store the data in 3 bytes for further processing. Two methods are shown for collecting the data, one using the PS/2 clock and the other using a sampled version with a much faster internal clock.

13

A Simple VGA Interface

Introduction

The Video Graphics Array (VGA) interface is common to most modern computer displays and is based on a pixel map, color planes and horizontal and vertical sync signals. A VGA monitor has three color signals (red, green and blue) that set one of these colors on or off on the screen. The intensity of each of those colors sets the final color seen on the display. For example, if the red was fully on, but the blue and green off, then the color would be seen as a strong red. Each analog intensity is defined by a two bit digital word for each color (e.g. red0 and red1) that are connected to a simple digital to analog converter to obtain the correct output signal.

The resolution of the screen can vary from 480×320 up to much large screens, but a standard default size is 640×480 pixels. This is 480 lines of 640 pixels in each line, so the aspect ratio is 640/480 leading to the classic landscape layout of a conventional monitor screen.

The VGA image is controlled by two signals – horizontal sync and vertical sync. The horizontal sync marks the start and finish of a line of pixels with a negative pulse in each case. The actual image data is sent in a $25.17\,\mu s$ window in a $31.77\,\mu s$ space between the sync pulses. (The time that image data is *not* sent is where the image is defined as a blank space and the image is dark.) The vertical sync is similar to the horizontal sync except that in this case the negative pulse mark the start and finish of each frame as a whole and the time for the frame (image as a whole) takes place in a $15.25\,ms$ window in the space between pulses, which is $16.784\,ms$.

There are some constraints about the spacing of the data between pulses which will be considered later in this chapter, but it is clear that the key to a correct VGA output is the accurate definition of timing and data by the VHDL.

Basic pixel timing

If there is a space of $25.17\,\mu s$ to handle all of the required pixels, then some basic calculations needs to be carried out to make sure that the Field Programmable Gate Arrays (FPGA) can display the correct data in the time available. For example, if we have a 640×480 VGA system, then that means that 640 pixels must be sent to the monitor in $25.17\,\mu s$. Doing the simple calculation shows that for each pixel we need $25.17\,\mu s/640 = 39.328\,ns$ per pixel. If our clock frequency is 100 MHz on the FPGA then that gives a minimum clock period of 10 ns, so this can be achieved with a relatively standard FPGA.

Image handling

Clearly it is not sensible to use an integrated image system on the FPGA, but rather it makes much more sense to store the image in memory (Random Access Memory (RAM)) and retrieve it frame by frame. Therefore as well as the basic VGA interface it makes a lot of sense for the images to be moved around in memory and therefore using the same basic RAM interface as defined previously is sensible. Therefore, as well as the VGA interface pins, our VGA handler should include a RAM interface.

VGA interface VHDL

The first stage in defining the VHDL for the VGA driver is to create a VHDL entity that has the global clock and reset, the VGA output pins and a memory interface. The outline VHDL entity is therefore given below:

```
Library ieee;
Use ieee.std_logic_1164.all;
Entity vga is
  Port (
        Clk : IN std_logic;
        Nrst : IN std_logic;
```

```
                    Hsync : OUT std_logic;
                    Vsync : OUT std_logic;
                    Red : OUT std_logic_vector (1 downto 0);
                    Green : OUT std_logic_vector (1 downto 0);
                    Blue : OUT std_logic_vector (1 downto 0);
                    Address : OUT (std_logic_vector (15 downto 0);
                    Data : IN (std_logic_vector (7 downto 0);
                    RAM_en : OUT std_logic;
                    RAM_oe : OUT std_logic;
                    RAM_wr : OUT std_logic
            );
        End entity vga;

        Architecture core of vga is
            -- VGA internal signals go here
        Begin
            -- VGA Interface core goes here
        End architecture core;
```

The architecture contains a number of processes, with internal signals that manage the transfer of pixel data from memory to the screen. As can be seen from the entity, the data comes back from the memory in 8 bit blocks and we require 3×2 bits for each pixel and so when the data is returned, each memory byte will contain the data for a single pixel. In this example, as we are using a 640×480 pixel image, this will therefore require a memory that is 307 200 bytes in size as a minimum. To put this in perspective, this means that using a raw memory approach we can put three frames per megabyte. In practice, of course, we would use a form of image compression (such as JPEG for photographic images), but this is beyond the scope of this book.

We can therefore use a simple process to obtain the current pixel of data from memory as follows:

```
Mem_read : process ( pclk, nrst ) is
        signal current_address : unsigned (16 downto 0);
Begin
        If nrst = '0' then
                Pixelcount <= 0;
                Current_address <= 0;
        Else
                If rising_edge(pclk) then
                    Current_address <= current_address + 1;
                    Address <= std_logic_vector
                      (current_address);
                    Pixel_data <= data;
                End if;
        End if;
End process;
```

This process returns the current value of the pixel data into a signal called pixel_data which is declared at the architecture level:

```
signal pixel_data : std_logic_vector ( 7 downto 0 );
```

This has the red, green and blue data defined in lowest 6 bits of the 8 bit data word with the indexes, respectively, of 0–1, 2–3, 4–5.

Horizontal sync

The next key process is the timing of the horizontal and vertical sync pulses, and the blanking intervals. The line timing for VGA is 31 770 ns per line with a window for displaying the data of 25 170 ns. If the FPGA is running at 100 MHz (period of 10 ns) then this means that each line requires 3177 clock cycles with 2517 for each line of pixel data, with 660 pulses in total for blanking (330 at either side). This also means that for a 640 pixel wide line, 39.3 ns are required for each pixel. We could round this up to 4 clock cycles per pixel. As you may have noticed, for the pixel retrieval we have a new internal clock signal called pclk, and we can create a process that generates the appropriate pixel clock (pclk) with this timing in place.

With this slightly elongated window, the blanking pulses must therefore reduce to 617 clock cycles and this means 308 before and 309 after the display window.

The horizontal sync pulse, on the other hand takes place between 26 110 ns and 29 880 ns of the overall interval. This is 189 clock pulse less than the overall line time, and so the horizontal sync pulse go low after 94 clock cycles and then at the end must return high 95 clock cycles prior to the end of the line. The difference between the outside and inside timings for the horizontal sync pulse is 377 clock cycles and so the sync pulse must return high 94 + 188 clock cycles and then return low 95 + 189 prior to the end of the window.

Thus the horizontal sync has the following basic behavior:

Clock Cycle	Value
0	1
94	0
282	1
2893	0
3082	1

And this can be implemented using a process with a simple counter:

```
Hsync_counter : process ( clk, nrst ) is
      Hcount : unsigned ( 11 downto 0 );
Begin
      If nrst = '0' then
            Hcount <= 0;
            Hsync <= '1';
      Else
            If hcount > and hcount < 2988 then
                  hsync <= '0';
            else
                  hsync <= '1';
            End if;
            If hcount < 3177 then
                  Hcount <= hcount + 1;
            Else
                  Hcount <= 0;
            End if;
      End if;
End process;
```

Vertical sync

The horizontal sync process manages the individual pixels in a line, and the vertical sync does the same for the lines as a whole to create the image. The period of a frame (containing all the lines) is defined as 16 784 000 ns. Within this timescale, the lines of the image are displayed (within 15 250 000 ns), then the vertical blanking interval is defined (up to the whole frame period of 16 784 000 ns) and finally the vertical sync pulse is defined as 1 until 15 700 000 ns at which time it goes to zero, returning to 1 at 15 764 000 ns.

Clearly it would not be sensible to define a clock of 10 ns for these calculations, so the largest common divisor is a clock of 2 μs, so we can divide down the system clock by 2000 to get a vertical sync clock of 2 μs to simplify and make the design as compact as possible.

```
Clk_div : process ( clk, nrst ) is
Begin
      If nrst = '0' then
            Count <= 0;
            Vclk <= '0';
      Else
            If count = 1999 then
                  Count <= 0;
                  Vclk <= not vclk;
```

```
                Else
                        Count <= count + 1;
                End if;
           End if;
     End process;
```

Where the vertical sync clock (vclk) is defined as a std_logic signal in the architecture. This can then be used to control the vsync pulses in a second process that now waits for the vertical sync derived clock:

```
Vsync_timing : process (vclk) is
Begin
     If nrst = '0' then
           Vcount <= 0;
     Else
           If vcount>15700 and vcount < 15764 then
                Vsync <= '0';
           Else
                Vsync <= '1';
           End if;
           If vcount > 16784 then
                Vcount <= 0;
           Else
                Vcount <= vcount + 1;
           End if;
     End if;
End process;
```

Using this process, the vertical sync (frame synchronization) pulses are generated.

Horizontal and vertical blanking pulses

In addition to the basic horizontal and vertical sync pulse counters, we have to define a horizontal blanking pulse which sets the line data low after 25 170 ns (2517 clock cycles). This can be implemented as a simple counter in exactly the same way as the horizontal sync pulse and similarly for a vertical blanking pulse. The two processes to implement these are given in the following VHDL.

```
Hblank_counter : process ( clk, nrst ) is
     Hcount : unsigned ( 11 downto 0 );
Begin
     If nrst = '0' then
           Hcount <= 0;
           hblank <= '1';
```

```
            Else
                    if hcount > 2517 and hcount < 3177 then
                            hblank <= '0';
                    else
                            hblank <= '1';
                    End if;
                    If hcount < 3177 then
                            Hcount <= hcount + 1;
                    Else
                            Hcount <= 0;
                    End if;
            End if;
    End process;
    Vblank_timing : process (vclk) is
    Begin
            If nrst = '0' then
                    Vcount <= 0;
                    Vblank<='1';
            Else
                    If vcount > 15250 and vcount < 16784 then
                            vblank <= '0';
                    Else
                            vblank <= '1';
                    End if;
                    If vcount > 16784 then
                            Vblank <= 0;
                    Else
                            Vcount <= vcount + 1;
                    End if;
            End if;
    End process;
```

Calculating the correct pixel data

As we have seen previously, the data of reach pixel is retrieved from a memory location and this is obtained using the pixel clock (pclk). The pixel clock is simply a divided (by 4) version of the system clock and at each rising edge of this pclk signal, the next pixel data is obtained from the memory data stored in the signal called data and translated into the red, green and blue line signals. This is handled using the basic process given below:

```
Pixel_handler : process (pclk) is
Begin
        Red <= data(1 downto 0);
        Green <= data(3 downto 2);
        Blue <= data(5 downto 4);
End process;
```

This is a basic handler process that picks out the correct pixel data, but is does not include the video blanking signal and if this is included, then the simple VHDL changes slightly to this form:

```
Pixel_handler : process (pclk) is
        Blank : std_logic_vector (1 downto 0);
Begin
        Blank(0) <= hblank or vblank;
        Blank(1) <= hblank or vblank;
        Red <= data(1 downto 0) & blank;
        Green <= data(3 downto 2) & blank;
        Blue <= data(5 downto 4) & blank;
End process;
```

This is the final step and completes the basic VHDL VGA handler.

Summary

This chapter has introduced the basics of developing a simple VGA handler in VHDL. While it is a simplistic view of the process, hopefully it has shown how a simple VGA interface can be developed using not very complex VHDL and a building block approach. It is left to the reader to develop their own complete VGA routines for the specific monitor that they have using the techniques developed in this chapter as a basis.

Part 4
Optimizing Designs

In this part of the book we will introduce a number of 'advanced' topics. In the other parts of the book, the emphasis is on the 'what', however in this part is it more about the 'how'. How can we make designs synthesize? How can our designs be made smaller or faster? How can we interface to mixed signal systems in practice? How can we develop verifiable designs? All of these design challenges will be addressed in this part of the book.

14
Synthesis

Introduction

The original intention of VHDL was to have a design specification language for digital circuits. The main goal of the work was to have a design representation that could be simulated to test whether the specification was fit for purpose. When VHDL was standardized as IEEE Std 1076, the broader application of VHDL for not just simulation but as an integral part of the hardware design flow became possible.

The original method of designing digital circuits was primarily through the development of schematic-based designs, using gate libraries to effectively generate Register Transfer Logic (RTL) netlists directly from the schematics. This is clearly a reasonable technique when the designs are relatively small, however it quickly becomes apparent that for designs of any size this approach is simply not realistic for modern Field Programmable Gate Arrays (FPGAs) that require millions of gates.

EDA companies realized fairly early on in the VHDL development process that if there was a standard language that could represent a data flow and a control flow, then the potential existed for automatically generating the gate level VHDL from a higher level description, and RTL was the obvious place to start. RTL has the advantage of representing the data flow and control flow directly, and can be mapped easily onto standard gate level logic. The resulting synthesis software (such as the Design Compiler from Synopsys) quickly established an important role in the digital design flow for both ASIC and FPGA designs and have in fact provided to be the driving force in the explosion of productivity of digital designers. The modern high density designs would not be possible without RTL synthesis.

As such, modern day designers often simplify 'RTL synthesis' to just 'synthesis', however this is not the whole story. As designs have continued to get more complex, there has been a push to ever increasing behavioral synthesis however there is not the same support from the EDA industry for behavioral synthesis software.

VHDL supported in RTL synthesis

While VHDL is standardized, synthesis is not, and as such the VHDL that can be synthesized is a subset of the complete VHDL language. Another common problem for designers is the fact that different synthesis software packages will give different output results for the same input VHDL, even to the extent that some will synthesize and some will not under certain conditions.

There are some standard VHDL techniques that cannot be synthesized however and these are summarized in this chapter.

There are two types of unsupported elements in VHDL – those that will cause a synthesis failure and those that are ignored. The failure elements are in many respects easier to manage as the synthesis software will provide an error message. It is the 'ignored' elements that can be more insidious as they can obviously leave errors in the synthesized design that may not be picked up until the hardware is tested.

Initial conditions

VHDL supports the initial condition being set for signals and variables, however this is not physically realized. In practice the initial conditions in the synthesized design are random and so in a practical design a reset condition should always be defined using an external reset pin. This is because during synthesis, the initial conditions are ignored.

Concurrent edges

It is common to use a clock edge as a trigger for a model, so a simple VHDL model may have a process with VHDL something like this to wait for the rising edge of a clock:

```
Process (clk)
    If rising_edge(clk) then
        Q <= q;
    End if;
End process;
```

Or in a similar way:

```
Process (clk)
    If clk'event and clk = '1' then
        Q <= q;
    End if;
End process;
```

What is NOT valid is to have more than one rising edge as the trigger condition:

```
Process (clk1, clk2)
    If rising_edge(clk1) and rising_edge(clk2) then
        Q <= d;
    End if;
End process;
```

This would fail the synthesis.

Numeric types

Synthesis is only supported for numbers that have a finite range. For example, an integer type with an undefined range (infinite) is not supported by synthesis software. In general terms it is often required that designers specify the range of integers and other integer-based numbers prior to synthesis (such as signed or unsigned).

This can be a subtle restriction as vectors that have a number as the index, must have this number defined in advance, so busses cannot be of a variable size.

Floating point (real) numbers are generally not supported by synthesis software tools as they do not have floating point libraries defined.

Wait statements

Wait statements are only supported if the wait is of the form of an implied sensitivity list *and a specific value*. So, if the statement is something like:

```
Wait on clk = '1';
```

Then this is supported for synthesis. If the wait statement is dependent on a specific time delay then this is NOT supported for synthesis. For example a statement in VHDL such as this is not supported:

```
Wait for 10 ns;
```

Assertions

Assertions in any form are ignored by the synthesis software.

Loops

The for loop is a special case of the general loop mechanism in VHDL and synthesis requires that the range of the loop must be defined as a static value, globally. This means that you cannot use variables to define the range of the for loop 'on the fly' for synthesis.

If a while loop is implemented, then there has to be a wait statement in the loop somewhere – otherwise it becomes a potentially infinite loop.

Some interesting cases where synthesis may fail

Unfortunately, there are differences between synthesis software packages and so care must be taken to ensure interoperability between packages, particularly in multi-team designs or when using third party VHDL cores. The cores may have been synthesized using software different to the one you are using in your design flow, so the advertised 'synthesizable' core may not always be synthesizable for you, in your design flow.

As such it is usually a good idea to keep the VHDL as generic as possible and avoid using 'tricks' of a particular package if you plan to deliver IP cores or use different tools. This may lead to slightly less compact VHDL, but the reliability of the VHDL will be greater, and potential problems (which could cause significant delays later in the design process, particularly in an integration phase) can be avoided.

One case is the use of different trigger variables in a process. For example, if there is a clock and a reset signal, or a clock and an enable signal, it is tempting to combine the logic into one expression such as:

```
If (clk'event and clk = '1' and nrst = '1') then
    . . .
End if;
```

However, in some synthesis software this would cause an error. It is always preferable to separate these variables into nested if statements for three reasons: (1) the code will be more readable, (2) the

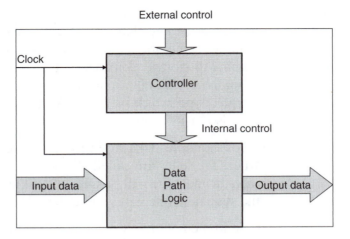

Figure 35
Synthesizable Digital
Circuit

chance of undefined logic states is reduced and (3) the synthesis software will not have a problem with your VHDL!

What is being synthesized?

Overall design structure

The basic approach for synthesizing digital circuits is to consider every design block as a combination of a controller and a data path. The controller is generally a Finite State Machine (FSM), clocked, and the data path is usually combinatorial logic, but there may also be storage in there and so a clock may also be required. The basic outline is shown in Figure 35.

Controller

The controller is producing the control signals for the data path logic and may also have external control signals, so there are both internal and external control signals in the general case. As this is a FSM, the design is synchronous and therefore is clocked and will generally have a reset condition.

The controller can be represented using a state diagram or bubble diagram. This shows each individual state and all the transitions between the states. The controller can be of two basic types: Moore (where the output of the state machine is purely dependent on the state variables) and Mealy (where the output can depend on the current state variable values AND the input values). The behavior of the state machine is represented by the state diagram (also sometimes called a state chart) as shown in Figure 36.

The technique for modeling FSMs will be covered later in this book, but the key elements to remember are that as this is a *Finite State Machine*, there are a *Finite* number of states, and hence the number of storage elements (D types) is implicit in this definition. Also, the VHDL allows the definition of the state names as an enumerated type, which makes the VHDL readable, easy to understand and also easily synthesizable.

For example, take a simple example of a two state machine, where the states are called ON and OFF. If the on off signal is low then the machine will be OFF and if the on off switch is high, then the state machine will go into the ON state.

To implement this simple state machine in VHDL, we can use a new type to represent the states:

```
Type states is (OFF, ON) ;
Signal current_state, next_state : states;
```

Notice that in the FSM VHDL we have defined both the current and the next state. The main part of the FSM can be easily implemented using a case statement in VHDL within a process that waits for changes in both the current_state signal and any external variables or control signals:

```
Process (current_state, onoff)
Begin
     Case current_state is
            When OFF =>
                    If onoff = '1' then
```

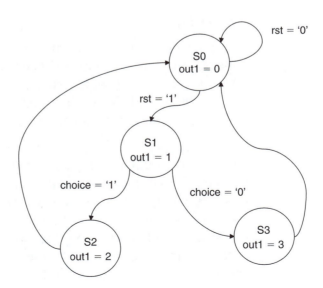

Figure 36
Basic State Machine

```
                        Next_state <= ON;
                End if;
        When ON =>
                If onoff = '0' then
                        Next_state <= OFF;
                End if;
        End case;
    End process;
```

Elsewhere in the architecture, the current_state needs to be assigned to the next state as follows:

```
Current_state <= next_state;
```

Data path

The data path logic is the logic (as the name suggests) to process the input data and generate the correct output data. The functionality of the data path logic will usually be divided into blocks and this offers the possibility of optimization for speed or area. For example, if area is not an issue, but speed is the primary concern, then a large design could be constructed to generate the output in potentially a single clock cycle. If the area is not an issue, but throughput is required, then pipelining could be used to maximize the overall data rates, although the individual latency may be high. Finally, if area is the critical factor, then single functional blocks can be used and registers used to store intermediate values and the same function applied repeatedly. Clearly this will be a lot slower, but potentially take a lot less space.

In the basic data path model there are blocks of combinational logic separated by registers. Clearly there are options for optimizing

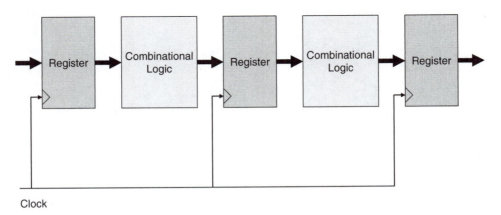

Figure 37
Data Path

the data flow by considering how best to move the data between the registers for speed or area optimization.

It is important to ensure that some simple rules are followed to ensure robust synthesis. The first is to make sure that each signal in the combinational block is defined for every cycle, in other words it is important not to leave undefined branches in case or if statements. If this occurs, then a memory latch is inferred and therefore a latch will be synthesized and as this is not linked to the global clock, unpredictable behavior can result.

Summary

This chapter has introduced the concept of synthesis, both from a designers point of view and also the implications of using certain types of VHDL with the intention of synthesizing it. The assumptions and limitations of the various approaches have been described and some sensible practical approaches to obtaining more robust designs defined.

15

Behavioral Modeling in VHDL

Introduction

There is a real need to abstract to a higher level in many designs to make the overall system level design easier. There is less need to worry about details of implementation at the system level if the design can be expressed behaviorally, especially if the synthesis method can handle any clock, partitioning or implementation issues automatically.

Furthermore, by using system level, or behavioral, analysis, decisions can be made early in the design process so that potentially costly mistakes can be avoided. Preliminary area and power estimates can be made and key performance specifications and architectural decisions can be made using this approach, without requiring to have detailed designs for every block.

How to go from RTL to behavioral VHDL

The abstraction from RTL (Register Transfer Level) VHDL to behavioral is straightforward in one sense, in that the VHDL is actually simpler. There is no need to ensure that correct clocking takes place, or that separate processes are implemented for different areas of the architecture, or even separate components instantiated.

It is useful to consider an example to illustrate this point by looking at the difference between the RTL and behavioral VHDL in an example such as a cross product multiplier. In this case we will demonstrate the RTL method and then show how to abstract to a behavioral model. First consider the specification for the model shown in Figure 38.

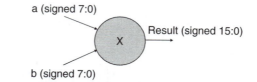

Figure 38
Cross Product
Multiplier
Specification

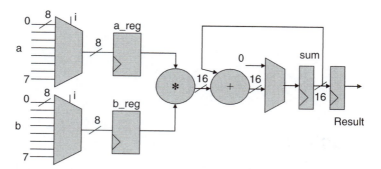

Figure 39
Data Path Model

This has the data path model as shown in Figure 39.

The first task is to define the types for the VHDL for the entity of the model and this is shown below. Notice that we have defined a new type sig8 that is a signed type and a vector based on this for the cross product multiplications.

```
library ieee;
Use ieee.std_logic_1164.all;
Use ieee.numeric_std.all;
Package cross_product_types is
    subtype sig8 is signed (7 downto 0);
    type sig8_vector is array
        (natural range<>) of sig8;
End package;
```

The entity can now be put together and this is shown below. Notice that for RTL we require both a clock and a reset.

```
library ieee;
use ieee.std_logic_1164.all;
use ieee.numeric_std.all;
use work.cross_product_types.all;

entity cross_product is
    port(
        a,b : in sig8_vector(0 to 7);
        clk, reset : in bit;
        result : out signed(15 downto 0)
    );
end entity cross_product;
```

The basic architecture can be set up that has the basic internal signals defined, and the processes will be explained separately.

```
architecture rtl of cross_product is
  signal I : unsigned (2 downto 0);
  signal ai, bi : sig8;
  signal product, add_in, sum, accumulator : signed (15
    downto 0);
begin
      control: process (clk)
      begin
        if clk'event and clk = '1' then
          if reset = '1' then
            i <= (others => '0');
          else
            i <= i + 1;
          end if;
        end if;
      end process;
      a_mux: ai <= a(i);
      b_mux <= bi <= b(i);
      multiply: product <= ai * bi;
      z_mux: add_in <= X"000" when i = 0 else
        accumulator;

      accumulate: process (clk)
      begin
        if clk'event and clk = '1' then
          accumulator <= sum;
        end if;
      end process;

      output : result <= accumulator;
  end;
```

Notice that there are two processes, one for the accumulation and the other to handle the multiplication. One important aspect is that it is not immediately obvious what is going on. Even in this simple model it is difficult to extract the key behavior of the state machine. In a complex controller it verges on the impossible unless the structure is well known and understood, which is an important lesson when using any kind of synthesis tool using VHDL or Verilog at any level.

Now reconsider using behavioral VHDL instead. The model uses the same packages and libraries as the RTL model, however notice that there is no need for an explicit clock or reset.

```
library ieee;
use ieee.std_logic_1164.all;
use ieee.numeric_std.all;
use work.cross_product_types.all;
```

```
entity cross_product is
   port(
      a,b : in sig8_vector(0 to 7);
      result : out signed(15 downto 0)
   );
end entity cross_product;
```

In this model, the architecture becomes much simpler and can be modeled in a much more direct way than the RTL approach.

```
architecture behav of cross_product is
begin

  process
    variable sum : signed(15 downto 0);
  begin
    sum := to_signed(0,16);
    for i in 0 to 7 loop
      sum := sum + a(i) * b(i);
    end loop;
    result <= sum;
    wait for 100 ns;
  end process;

end architecture;
```

Notice that it is much easier to observe the functionality of the model and also the behavior can be debugged more simply than in the RTL model. The design is obvious, the code is readable and the function is easily ascertained. Note that there is no explicit controller, the synthesis mechanism will define the appropriate mechanism. Also notice that the model is defined with a single process. The synthesis mechanism will partition the design depending on the optimization constraints specified.

Note the wait statement. This introduces an implicit clock delay into the system. Obviously this will depend on the clock mechanism used in reality. There is also an implied reset. If an explicit clock is required then use a wait until rising_edge (clk) or similar approach, but retaining the behavioral nature of the model.

Consider a another useful example: a Finite Impulse Response (FIR) filter. Ignoring the entity and declarations, how can we model an ideal FIR filter behaviorally in VHDL?

The specification of the interface is as follows:

```
Input : signed (15 downto 0)
Output : signed(15 downto 0)
Coefficients : array(natural range<>) of integer...
```

And the resulting VHDL code would be something like the following:

```
for I in samples'right downto 1 loop
    samples(I) := samples(I-1);
end loop
samples(0) := input;

sum := to_signed(0,32);
for j in 0 to samples'right loop
    sum := sum + (to_signed(coeffs(j),16) *
        samples(j));
end loop;

output <= sum(30 downto 15);
wait for 1 us;
```

This is easily parameterized, modified and clear to understand.

Summary

Behavioral VHDL is a useful technique for both initial design ideas and also the starting point for an RTL design. It is important to remember, however, that quite a lot of behavioral VHDL cannot be synthesized and is therefore purely for conceptual design or use in test benches. In order to make this a practically useful design tool, the designer can take advantage of the ability of VHDL to have numerous architectures and by using the same test bench validate the RTL against the behavioral model to ensure correctness.

In summary, we can use behavioral modeling early with high impact to:

- carry out fast functional simulation,
- make performance criteria/design trade-offs,
- investigate non-linear effects,
- look at implementation issues,
- carry out topology evaluation.

16
Design Optimization

Introduction

The area of design optimization is where the performance of a design can be made drastically better than an initial naïve implementation. Before discussing details of how to make the designs optimal for the individual goals of speed, area and power, the 'big three' for design optimization generally in digital design and particularly for Field Programmable Gate Arrays (FPGAs), it is useful to discuss some principles of what happens when we synthesize a function into hardware.

There are two main areas for optimization of the design when working with VHDL for FPGAs and these are in the optimization of the Register Transfer Level (RTL) VHDL which is leading to an optimal description of the VHDL in terms of logic expressions. The second key area is in the basic logic minimization prior to the mapping of low-level functions to the individual technology gates.

Techniques for logic optimization

There are two approaches to minimizing the logic in a design, one that maintains the hierarchy and the other that flattens it. Often a synthesis tool will allow the user to choose which option is required. Clearly the advantage of flattening a design is that the logic can be considered as a whole, whereas if the logic hierarchy is maintained, then there may be structural aspects of the design that will be of benefit to the behavior of the circuit as a whole.

The basic approach of the logic minimization is to reduce the logic equation set to a two level form (otherwise known as sum-of-products). The most common approach for simple designs is to use a Karnaugh map to show the input and output variables graphically and then produce an output expression that can provide the same outputs but using a smaller amount of logic than the original Boolean expressions.

For example, consider the basic 4 input Karnaugh map shown in Figure 40.

When a logic expression is described using a logic equation, we can select all valid outputs by circling all the required output '1's and this defines the basic logic behavior. The basic technique is to make the circles as large as possible to encompass as many output '1's with as few input variables as possible. For example, if a basic logic equation was defined as $Z = A \cdot B \cdot \overline{C} + \overline{A} \cdot B \cdot D + \overline{A} \cdot B \cdot D$, then the resulting Karnaugh map would be as shown in Figure 41.

Currently, with this basic implementation this would require 3, 3 input AND gates, a 3 input OR gate and several inverters. We can

CD \ AB	00	01	11	10
00	Z_0	Z_4	Z_8	Z_{12}
01	Z_1	Z_5	Z_9	Z_{13}
11	Z_2	Z_6	Z_{10}	Z_{14}
10	Z_3	Z_7	Z_{11}	Z_{15}

Figure 40
Basic 4 Input Karnaugh Map

CD \ AB	00	01	11	10
00	0	0	1	0
01	0	1	1	0
11	0	1	0	0
10	0	0	0	0

Figure 41
Specific Karnaugh Map Example

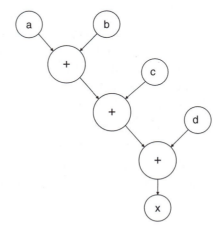

Figure 42
Functions Identified
on Karnaugh Map

Figure 43
Naïve Dataflow
Diagram for Addition

see from the Karnaugh map however that if we define only two of those logic functions, that there is redundancy in the original definition, and we can reduce this to the same output for two logic combinations of the input in Figure 42.

We could therefore define this model using the simplified expression defined as $Z = A{\cdot}B{\cdot}\overline{C} + \overline{A}{\cdot}B{\cdot}D$ which has clearly reduced the size of the logic by 1, 3 input AND gate and the OR gate has reduced to a 2 input gate.

Improving performance

Consider a simple example of an addition $x = a + b + c + d$, where all the variables are digital words. We could implement this using adders taking two numbers at a time and then adding the answer to the next input. This would give the following data flow diagram in Figure 43.

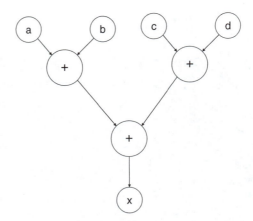

Figure 44
Reduced Cycle
Implementation

This implementation requires 3 adders and takes 3 cycles to get the answer. If we were more systematic with the same resources, we could reduce this to two cycles by adopting a different structure shown in Figure 44.

This is a classic case of an expression tree being reduced so that the control path can take fewer cycles, but achieving the same data path result. We could also the visage the case where we only use a single addition block, but use registers to store the intermediate sums and then 'pipeline' the sums until we complete the expression. This would potentially take the longest, however would result in the smallest area requirement – as there would only be the need for a single addition block (however of course this would be a trade-off with an increased number of registers).

Critical path analysis

Another approach to logic optimization is to analyze the critical path through a design from a timing perspective. This is often carried out automatically by the synthesis software, for example the Synopsys Design Compiler software automatically generates a synthesized schematic that highlights the critical path through the design for timing and as such the designer can concentrate their efforts on that area of the design to improve the overall throughput in that case shown in Figure 45.

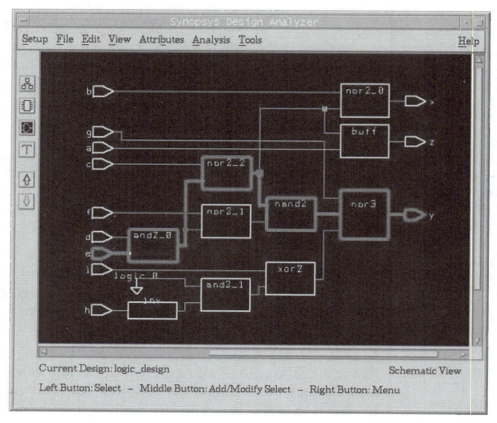

Figure 45
Critical Path Analysis

Summary

This chapter has discussed some techniques for improving the performance of VHDL designs on FPGAs and how they work. Much of the actual optimization is taken care of in the synthesis software, however it is useful to understand the processes involved so that of a specific target is required for optimization this can be achieved in a reasonable time in a controlled manner.

17
VHDL-AMS

Introduction

With the increasingly high level of system integration it is becoming necessary to model not only electronic behavior of systems, but also interfaces to 'real-world' applications and the detailed physical behavior of elements of the system in question. The emergence of standard languages such as VHDL-AMS has made it possible to now describe a variety of physical systems using a single design approach and simulate a complete system. Application areas where this is becoming increasingly important include mixed-signal electronics, electro-magnetic interfaces, integrated thermal modeling, electro-mechanical and mechanical systems (including micro-electro-mechanical systems, MEMS), fluidics (including hydraulics and micro-fluidics), power electronics with digital control and sensors of various kinds.

In this chapter, we will show how the behavioral modeling of multiple energy domains is achieved using VHDL-AMS, demonstrating with the use of examples how the interactions between domains takes place, and provide an insight into design techniques for a variety of these disciplines. The basic framework is described, showing how standard packages can define a coherent basis for a wide range of models, and specific examples used to illustrate the practical details of such an approach. Examples such as integrated simulation of power electronics systems including electrical, magnetic and thermal effects, mixed-domain electronics and mechanical systems are presented to demonstrate the key concepts involved in multiple energy domain behavioral modeling.

The basic approach for modeling devices in VHDL-AMS is to define a model entity and architecture(s). The model entity defines the interface of the model to the system and includes connection points and parameters. A number of architectures can be associated with an entity to describe the model behavior such as a behavioral or a physical level description. A complete model consists of a single entity combined with a single architecture. The domain or technology type of the model is defined by the type of terminal used in the entity declaration of the ports. The IEEE Std 1076.1.1 defines standard types for multiple energy domains including electrical, thermal, magnetic, mechanical and radiant systems. Within the architecture of the model, each energy domain type has a defined set of through and across variables (in the electrical domain these are voltage and current, respectively) that can be used to define the relationship between the model interface pins and the internal behavior of the model.

In the 'conventional' electronics arena, the nature of the VHDL-AMS language is designed to support 'mixed-signal' systems (containing digital elements, analog elements and the boundary between them) with a focus on IC design. Where the strengths of the VHDL-AMS language have really become apparent, however, is in the multi-disciplinary areas of mechatronic and MEMS. In this chapter, I have highlighted several interesting examples that illustrate the strengths of this modeling approach, with emphasis on multiple domain simulations.

Introduction to VHDL-AMS

VHDL-AMS is a set of analog extensions to standard digital VHDL to allow mixed-signal modeling of systems. The VHDL-AMS language was approved as IEEE Std 1076.1 in 1999; however, it is important to note that IEEE 1076.1-1999 encompasses the complete digital VHDL 1076 standard and is not a subset.

The standard does not specify any libraries for analog disciplines (e.g. electrical, mechanical, etc.). This is a separate exercise and is covered by a subset working group IEEE 1076.1.1, which was released as an IEEE Std 1076.1.1 in 2004.

In order to put the extensions into context it is useful to show the scope of VHDL, and then VHDL-AMS alongside it and this is shown in Figure 46.

The key extensions for VHDL-AMS is the ability to look upward to transfer functions (behavioural and in the Laplace domain) and downwards to differential equations at the circuit level.

Level	Content
System	Transfer Function
System	Specification
Chip	Algorithms
Register	Truth Tables State Tables
Logic	Boolean Equations
Circuit	Differential Equations

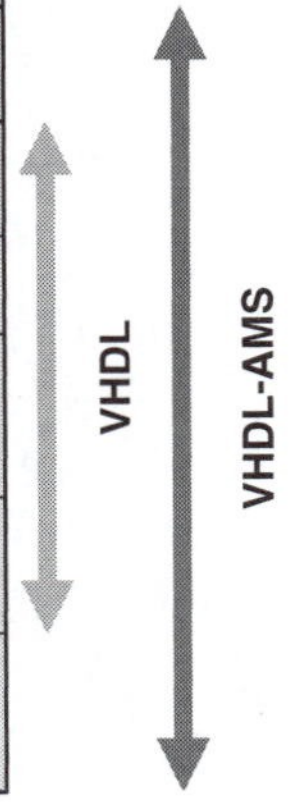

Figure 46
Scope of VHDL-AMS

The extensions to VHDL for VHDL-AMS can be summarized as follows:

1. A new type of ports called TERMINALS – basically analog pins.
2. A new type of TYPE called a NATURE that defines the relationship between analog pins and variables.
3. A new type of variable called a QUANTITY that is an analog variable.
4. A new type of variable assignment that is used to define analog equations that are solved simultaneously.
5. Differential equation operators for derivative ('DOT) and integration ('INTEG) with respect to time.
6. IF statements for equations (IF USE).
7. Break statement to initialize the non-linear solver.
8. STEP LIMIT control for limiting the analog time step in the solver.

Analog pins: TERMINALS

In order to define analog pins in VHDL-AMS we need to use the TERMINAL keyword in a standard entity PORT declaration. For example, if we have a two pins device that has two analog pins (of type electrical, more on this later), then the entity would have the basic form as shown below:

```
LIBRARY IEEE;
USE IEEE.ELECTRICAL_SYSTEMS.ALL;
ENTITY model IS
GENERIC();
PORT(
     TERMINAL p : electrical;
     TERMINAL m : electrical
     );
END ENTITY;
```

Notice that as the VHDL-AMS extensions are defined as an IEEE standard, then the use of a standard library such as electrical pins requires the use of the electrical_systems.all; packages from the IEEE library.

Notice that the pins do not have a direction assigned as analog pins are part of a conserved energy system and are therefore solved simultaneously.

Mixed-domain modeling

In order to use standard models, there has to be a framework for terminals and variables which is where the standard packages are used. There is a complete IEEE Std (1076.1.1) which defines the standard packages in their entirety; however, it is useful to look at a simplified package (electrical systems in this case) to see how the package is put together.

For example, electrical systems models need to be able to handle several key aspects:

- Electrical connection points

- Electrical 'through' variables (i.e. current)

- Electrical 'across' variables (i.e. voltages)

The electrical systems 'package' needs to encompass these elements.

First, the basic subtypes need to be defined. In ALL the analog systems and types, the basic underlying VHDL type is always

REAL, and so the voltage and current must be defined as subtypes of REAL:

```
Subtype voltage is REAL;
Subtype current is REAL;
```

Notice that there is no automatic unit assignment for either, but this is handled separately by the UNIT and SYMBOL attributes in IEEE Std 1076.1.1. For example, for voltage the unit is defined as 'Volt' and the symbol is defined as 'V'.

The remainder of the basic electrical type definition then links these subtypes to the through and across variable of the type, respectively:

```
PACKAGE electrical_system IS
    SUBTYPE voltage IS real;
    SUBTYPE current IS real;
    NATURE electrical IS
        voltage ACROSS
        current THROUGH
        ground REFERENCE;
END PACKAGE electrical_system;
```

Analog variables: quantities

Quantities are purely analog variables and can be defined in one of three ways. Free quantities are simply analog variables that do not have a relationship with a conserved energy system. Branch quantities have a direct relationship between one or more analog terminals and finally source quantities are used to define special source functions (such as AC sources or noise sources).

For example to define a simple analog variable called x, that is a voltage (but not related directly to an electrical connection (TERMINAL), then the following VHDL could be used:

```
QUANTITY x : voltage;
```

On the other hand, a branch between two electrical pins has a through variable (current) and an across variable (voltage) and this requires a 'branch' quantity so that the complete description can be solved simultaneously. For example, the complete quantity declaration for the voltage (v) and current (i) of a component between two pins (p & m) could be defined as:

```
QUANTITY v across i through p to m;
```

Simultaneous equations in VHDL-AMS

In VHDL-AMS the equations are analog and solved simultaneously, which is in contrast to signals that are solved concurrently using logic techniques and variables which are evaluated sequentially. For example in VHDL-AMS to solve an equation use the '==' operator:

```
Y == x**2;
```

Where both Y and X have to be defined as real numbers (quantities or other VHDL variable types).

A VHDL-AMS example

A DC voltage source

In order to illustrate some of these basic concepts consider a simple example of a DC voltage source. This has two electrical pins p & m, and a single parameter dc_value that is used to define the output voltage of the source (Figure 47).

This can be modeled in VHDL-AMS in two parts, the entity and architecture. First, consider the entity. This has two electrical pins, so we need to use the ieee.electrical_systems.all; package and therefore the ports are to be declared as TERMINALS. Also the generic de_value must be defined as a real number with the default value also defined as a real number (e.g. 1.0):

```
LIBRARY IEEE;
USE IEEE.ELECTRICAL_SYSTEMS.ALL;
ENTITY v_dc IS
GENERIC(
    dc_value : real := 1.0);
PORT(
    TERMINAL p : electrical;
    TERMINAL m : electrical
    );
END ENTITY;
```

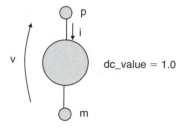

Figure 47
Basic Voltage Source

The architecture must define the quantities for voltage and current through the source and then link those to the terminal pin names. Also, the output equation of the source must be modeled as an analog equation in VHDL-AMS using the '==' operator to implement the function v = dc_value:

```
ARCHITECTURE simple OF v_dc IS
     QUANTITY v ACROSS I THROUGH p TO m;
BEGIN
     v == dc_value;
END ARCHITECTURE simple;
```

Resistor

In the case of the resistor, the basic entity is very similar to the voltage source with two electrical pins p & m with a single generic, this time for the nominal resistance rnom (Figure 48).

This can be modeled in VHDL-AMS in two parts, the entity and architecture. First consider the entity. This has two electrical pins, so we need to use the ieee.electrical_systems.all; package and therefore the ports are to be declared as TERMINALS. Also the generic rnom must be defined as a real number with the default value also defined as a real number (e.g. 1000.0):

```
LIBRARY IEEE;
USE IEEE.ELECTRICAL_SYSTEMS.ALL;
ENTITY resistor IS
GENERIC(
     rnom : real := 1000.0);
PORT(
     TERMINAL p : electrical;
     TERMINAL m : electrical
     );
END ENTITY;
```

The architecture must define the quantities for voltage and current through the resistor and then link those to the terminal pin names.

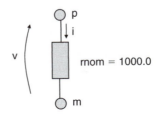

Figure 48
VHDL-AMS Resistor
Symbol

Also, the output equation of the resistor must be modeled as an analog equation in VHDL-AMS using the '==' operator to implement the function v = I * rnom:

```
ARCHITECTURE simple OF resistor IS
    QUANTITY v ACROSS I THROUGH p TO m;
BEGIN
    v == I * rnom;
END ARCHITECTURE simple;
```

Differential equations in VHDL-AMS

VHDL-AMS also allows the modeling of linear differential equations using the two differential operators:

1. 'DOT (Differentiate the variable with respect to time)

2. 'INTEG (Integrate the variable with respect to time)

We can illustrate this by taking two examples, a capacitor and an inductor. First, consider the basic equation of a capacitor:

$$i = C \frac{dV}{dt}$$

Using a similar model structure as the resistor, we can define a model entity and architecture, but what about the equation? In VHDL-AMS, the 'DOT operator is used on the voltage to represent the differentiation as follows:

```
i == c*v'DOT;
```

Therefore, a complete capacitor model in VHDL-AMS could be implemented as follows:

```
LIBRARY IEEE;
USE IEEE.ELECTRICAL_SYSTEMS.ALL;
ENTITY capacitor IS
GENERIC(
    cap : real := 1.0e-9);
PORT(
    TERMINAL p : electrical;
    TERMINAL m : electrical
    );
END ENTITY;
```

```
ARCHITECTURE simple OF capacitor IS
    QUANTITY v ACROSS I THROUGH p TO m;
BEGIN
    I == cap * v'DOT;
END ARCHITECTURE simple;
```

What about an inductor? The basic equation for an inductor is given below:

$$i = \frac{1}{L} \int v \, dt$$

Obviously, the most direct way to implement this equation would be to use the 'INTEG operator, however care should be taken with the integration operator.

Obviously, the initial condition must be considered and in addition different implementations can occur across simulators. However, the resulting implementation in its simplest form could be as follows:

```
LIBRARY IEEE;
USE IEEE.ELECTRICAL_SYSTEMS.ALL;
ENTITY inductor IS
GENERIC(
    ind : real := 1.0e-9);
PORT(
    TERMINAL p : electrical;
    TERMINAL m : electrical
    );
END ENTITY;

ARCHITECTURE simple OF inductor IS
    QUANTITY v ACROSS I THROUGH p TO m;
BEGIN
    I == (1.0/ind) * v'INTEG;
END ARCHITECTURE simple;
```

Mixed-signal modeling with VHDL-AMS

Most design engineers are familiar with the concepts of 'digital' or 'analog' modeling; however, a true understanding of 'mixed-signal' modeling is often lacking. In order to explain the term 'mixed-signal modeling', it is necessary to review what we mean by analog and digital modeling first. First, consider digital modeling techniques.

Digital systems can be modeled using digital gates or events. This is a fast way of simulating digital systems structurally and is based on VHDL or Verilog gate level models. Digital simulation with digital computers relies on an event-based approach, so rather than solve differential equations, events are scheduled at certain points in time, with discrete changes in level. The resolution of multiple events and connections is achieved using logic methods. The digital models are usually gates, or logic based and the resulting simulation waveforms are of fixed, predefined levels (such as '0' or '1'). Also, 'instantaneous' changes can take place, that is the state can change from '0' to '1' with zero risetime.

In the analog world, in contrast, the lowest level of detail in practical electrical system design is the use of analog equation models in an analog simulator – the benchmark of this approach is historically the SPICE simulator. In many cases the circuit is extracted in the form of a netlist. The netlist is a list of the components in the design, their connection points and any parameters (such as length, width or scaling) that customize the individual devices.

Each device is modeled using non-linear differential equations that must be solved using a Newton–Raphson type approach. This approach can be very accurate, but is also fraught with problems such as:

- *Convergence*: If the model does not converge, then the simulation will not give any meaningful result or fail altogether.

- *Oscillation*: If there are discontinuities, the solution may be impossible to find.

- *Time*: The simulations can take hours to complete, days for large designs with detailed device models.

In the analog domain the Newton–Raphson approach is generally used to find a solution which relies on calculating the derivatives as well as the function value to obtain the next solution. The basic Newton–Raphson method for non-linear equations is defined as:

$$x_{n+1} = x_n - \frac{F(x_n)}{F'(x_n)}$$

$F(x_n)$ and $F'(x_n)$ must be explicitly known and coded into the simulator (for SPICE) and this gives an approximate solution to the exact problem. For VHDL-AMS simulators the derivatives must be estimated using a Secant method (or similar) (Figure 49).

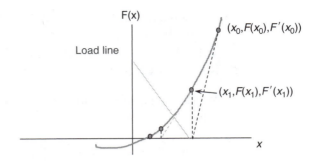

F(x)

Load line

$(x_0, F(x_0), F'(x_0))$

$(x_1, F(x_1), F'(x_1))$

x

Figure 49
Newton–Raphson
Method

So given these diametrically opposed methods, how can we put them together? What about mixed-signal systems? In these cases, there is a mixture of continuous analog variables and digital events. The models need to be able to represent the boundaries and transitions between these different domains effectively and efficiently. The basic mechanism to checking if an analog variable crosses a threshold is to use the ABOVE operator in VHDL-AMS.

For example, to check if a voltage 'vin' is above 1.0 V, the following VHDL-AMS could be used:

```
if ( vin'above(1.0) ) then
    flag <= true;
end if;
```

This can be extended to use parameters in the model – say a threshold voltage parameter (vth) – defined previously as a generic or constant:

```
if ( vin'above(vth) ) then
    flag <= true;
end if;
```

Notice that flag is a signal and is therefore able to be used in the sensitivity list to a process enabling digital behavior to be triggered when the threshold is crossed. If the opposite condition is required, that is BELOW the threshold, then the condition is simply inverted using the NOT operator:

```
if ( NOT vin'above(vth) ) then
    flag <= true;
end if;
```

The digital-to-analog interface is slightly more complex than the analog-to-digital interface inasmuch as the output variable needs to be controlled in the analog domain.

When a digital event changes (this can be easily monitored by a sensitivity list in a process) the analog variable needs to have the correct value and the correct rate of change. To achieve this we use the RAMP attribute in VHDL-AMS.

Consider a simple example of a digital-logic-to-analog-voltage interface:

- When Din = '1' Vout = 5 V

- When Din = '0' Vout = 0 V

This can be implemented using VHDL-AMS as follows:

```
process (din) :
begin
    if ( din = '1' ) then
        vdin = 5.0;
    else
        vdin = 0.0;
    end if;
end process;
vout == vdin;
```

Clearly, there will be problems with this simplistic interface as the transition of vout will be instantaneous – causing potential convergence problems. The technique to solve this problem is to introduce a RAMP on the definition of the value of vout with a transition time to change continuously from one value to another:

```
vout == dvin'RAMP(tt)
```

Where tt (the transition time) is defined as a real number (e.g. tt : real : = 1.0e − 9;).

An alternative to the specific transition time definition is to limit the slew rate using the SLEW operator. The technique to solve this problem is to introduce a slew rate definition on the definition of the value of vout with a transition time to change continuously from one value to another:

```
vout == dvin'SLEW(max_slew_rate)
```

Where max_slew_rate is defined as a real number (e.g. max_slew_rate : real : = 1.0e6;).

A basic switch model

Consider a simple digitally controlled switch that has the following characteristics:

- Digital control input (d)

- Two electrical terminals (p & m)

- On resistance (Ron)

- Off resistance (Roff)

- Turn on time (Ton)

- Turn off time (Toff)

Using this simple outline a basic switch model can be created in VHDL-AMS. The entity is given below:

```
USE ieee.electrical_system.ALL;
USE ieee.std_logic_1164.ALL;
ENTITY switch IS
    GENERIC ( ron : real := 0.1; — On resistance
            roff : real := 1.0e6; — Off resistance
            ton : real := 1.0e-6;   — turn on time
            toff : real := 1.0e-6); — turn off time
    PORT (
    d : IN std_logic;
    TERMINAL p,m : electrical);
END ENTITY switch;
```

The basic structure of the architecture requires that the voltage and current across the terminals of the switch be dependent on the effective resistance of the switch (reff):

```
ARCHITECTURE simple OF switch IS
    QUANTITY v ACROSS i THROUGH p TO m;
    QUANTITY reff : real;
    SIGNAL r_eff : real := roff;
BEGIN
    PROCESS (d)
    BEGIN
        ...
    END;

    i = v / reff;
END;
```

The process waits for changes on the input digital signal (d) and schedules a signal r_eff to take the value of the effective resistance (ron or roff) depending on the logic value of the input signal. The VHDL for this functionality is shown below:

```
PROCESS (d)
   BEGIN
       if ( d = '1' ) then
               r_eff <= ron;
       else
               r_eff <= roff;
       end if;
   END;
```

When the signal r_eff changes, then this must be linked to the analog quantity reff using the ramp function. Previously we showed how the ramp could define a risetime, but in fact it can also define a falltime. Implementing this in the switch model architecture we get the following VHDL-AMS:

```
reff == r_eff'RAMP ( ton, toff );
i == v / reff;
```

The complete VHDL-AMS model for the switch is given below:

```
ARCHITECTURE simple OF switch IS
    QUANTITY v ACROSS i THROUGH p TO m;
    QUANTITY reff : real;
    SIGNAL r_eff : real := roff;
BEGIN
    PROCESS (d)
          BEGIN
                  if ( d = '1' ) then
                          r_eff <= ron;
                  else
                          r_eff <= roff;
                  end if;
          END PROCESS;

          reff == r_eff'RAMP ( ton, toff );
          i == v / reff;
    END;
```

Basic VHDL-AMS comparator model

Consider a simple comparator that has two electrical inputs (p & m), an electrical ground (gnd) and a digital output (d). The comparator has a digital output of '1' when p is greater than m and '0' otherwise (Figure 50).

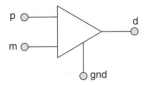

Figure 50
Comparator

The entity defines the terminals (p, m, gnd), digital output (d), input hysteresis (hys) and the propagation delay (td):

```
USE ieee.electrical_system.ALL;
USE ieee.std_logic_1164.ALL;
ENTITY comparator IS
    GENERIC (
td : time := 10 ns;
hys : real := 1.0e-6);
    PORT (
d : OUT std_logic := '0';
TERMINAL p,m,gnd : electrical);
END ENTITY comparator;
```

The first step in the architecture is to define the input voltage and basic process structure:

```
architecture simple of comparator is
 quantity vin across p to m;
begin
 p1 : process
        constant vh : real := ABS(hys)/2.0;
        constant vl : real := -ABS(hys)/2.0;
 begin
        ...
        wait on vin'above(vh), vin'above(vl);
 end process;
end architecture simple;
```

The quantity vin is defined as the voltage across the input pins p and m:

```
quantity vin across p to m;
```

Notice that no current is defined, assumed to be zero, so there is no input current to the comparator. Also notice that there is no input voltage offset defined – this could be added as a refinement to the model later. The process defines the upper and lower thresholds (vh and vl) based on the hysteresis:

```
constant vh : real := ABS(hys)/2.0;
constant vl : real := -ABS(hys)/2.0;
```

The process then defines a wait statement checking vin for crossing either of those threshold values:

```
wait on vin'above(vh), vin'above(vl);
```

The final part of the process is to add the digital output logic state dependent on the threshold status of vin:

```
if vin'above(vh) then
      d <= '1' after td;
elsif not vin'above(vl) then
      d <= '0' after td;
end if;
```

The output state (d) is then scheduled after the delay time defined by td.

The completed architecture is shown below:

```
architecture simple of comparator is
 quantity vin across p to m;
begin
 p1 : process
      constant vh : real := ABS(hys)/2.0;
      constant vl : real := -ABS(hys)/2.0;
 begin
      if vin'above(vh) then
            d <= '1' after td;
      elsif not vin'above(vl) then
            d <= '0' after td;
      end if;
      wait on vin'above(vh), vin'above(vl);
 end process;
end architecture simple;
```

Multiple domain modeling

A final significant application area for VHDL-AMS has been the modeling of electro-mechanical systems, particularly micromachines (or MEMS). Exactly the same principles are used for these devices, with the mechanical domain models defined as required for the mechanical equations. It is worth noting that the mechanical models are divided into rotational (angular velocity and torque) and translational (force and distance) types. A typical simple example of a mixed-domain system is a motor, in this case a simple DC motor. Taking the standard motor equations as shown below, it can be seen that the parameter ke links the rotor speed to the electrical domain (back emf) and the parameter kt links the current to the torque:

$$V = L\frac{di}{dt} + iR + Ke\omega$$

$$T = Kti - J\frac{d\omega}{dt} - D\omega$$

This is implemented using the VHDL-AMS model shown below:

```
Library ieee;
use ieee.electrical_systems.all;
use ieee.mechanical_systems.all;

entity dc_motor is
    generic (kt : real;
             j : real;
             r : real;
             ke : real;
             d : real;
             l : real);
    port (terminal p, m : electrical;
          terminal rotor : rotational_v);
end entity dc_motor;

architecture behav of dc_motor is
    quantity w across t through rotor
  to rotational_v_ref;
    quantity v across i through p to m;
begin
  v == l*i'DOT + i*r + ke*w;
  t == i*kt - j*w'DOT - d*w;
end architecture behav;
```

Summary

It has become crucial for effective design of integrated systems, whether on a macro- or microscopic scale, to accurately predict the behavior of such systems prior to manufacture. Whether it is ensuring that sensors or actuators operate correctly, or integrated components such as magnetics also operate correctly, or analyzing the effect of parasitics and non-ideal effects such as temperature, losses and non-linearities, the requirement for multiple domain modeling has never been greater.

Now languages such as VHDL-AMS offer an effective and efficient route for engineers to describe these systems and effects, with the added benefit of standardization leading to interoperability and model exchange. The challenge for the EDA industry is to provide adequate simulation and particularly modeling tools to support engineering design.

The opportunity for Field Programmable Gate Array (FPGA) designers is to take advantage of this huge advance in modeling technology and use it to make sure that digital controllers and designs can operate effectively and robustly in real-world applications.

18

Design Optimization Example: DES

Introduction

Elsewhere in this book the basics of design optimization are discussed, but in general these are at a Register Transfer Level (RTL) level. The use of behavioral modeling has also been described, but in general the use of behavioral synthesis is still rarely used. In this chapter, the use of behavioral synthesis is investigated as an alternative to create optimal designs rather than using an RTL approach.

This chapter describes the experience of designing a Data Encryption Standard (DES) core in Electronic Code Book (ECB) mode using the MOODS behavioral synthesis system. The main objective was to write a high-level language description that was both readable and synthesizable. The secondary objective was to explore the area/delay design space of both single and triple DES. The designs were simulated using both the pre-synthesis (behavioral) and post-synthesis (RTL) VHDL, verifying that the outputs were not only the same, but also were the expected outputs defined in the test set.

The DES

The DES, usually referred to by the acronym DES, is a well-established encryption algorithm which was first standardized by the National Institute of Standards and Technology (NIST) in the 1980s. It is described in detail later in this book in the chapter on secure systems, so only the basic information about the algorithm is presented here.

While DES has largely been superceded by AES (Advanced Encryption Algorithm), it is now common to find the algorithm being used in triplicate – an algorithm known as triple DES or TDES for short. This algorithm uses the same DES core, but uses three passes with different keys. DES was designed to be small and fast, and the algorithm is mainly based on shuffling and substitution – there is very little computation involved – which makes it ideal for hardware implementation.

Moods

MOODS (**M**ultiple **O**bjective **O**ptimization in Control and **D**atapath **S**ynthesis) is a high-level behavioral synthesis suite developed at the University of Southampton. It takes as input behavioral VHDL and transforms this into structural VHDL that is behaviorally equivalent. MOODS uses optimization and design space exploration to obtain suitable RTL designs to meet designer's constraints and requirements.

The optimizer converts the behavioral VHDL into a form that can be described using a simple Dataflow Graph (DFG) which allows the control flow to be optimized. This is effectively a state machine that can be easily converted into RTL VHDL. The optimization of this with respect to area can be achieved by sharing data units (such as registers) using multiplexing and with respect to delay by combining data units to reduce the number of clock cycles required.

Initial design

Introduction

The overall structure of the DES algorithm is shown in Figure 51.

The core algorithm is repeated 16 times with a different subkey for each round. These subkeys are 48 bits long and are generated from the original 56-bit key. The algorithm was converted directly to VHDL using a functional decomposition style (i.e. functions were created to represent each equivalent function in DES).

Overall structure

The first stage in this design was to create an entity and an architecture with the required inputs and outputs and a single process

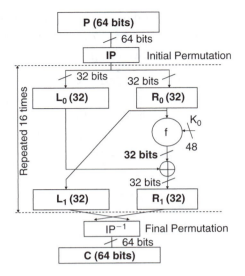

Figure 51
Overall Structure
of the DES Algorithm

containing the overall algorithm. This resulted in the VHDL outline below:

```
library ieee;
use ieee.std_logic_1164.all;
entity DES is
  port (
    plaintext : in std_logic_vector(1 to 64);
    key : in std_logic_vector(1 to 64);
    encrypt : in std_logic;
    go : in std_logic;
    ciphertext : out std_logic_vector(1 to 64);
    done : out std_logic
  );
end;

architecture behavior of DES is
  subtype vec56 is std_logic_vector(1 to 56);
  ...
  subtype vec64 is std_logic_vector(1 to 64);
begin
  process
  begin
    wait until go = '1';
    done <= '0';
    wait for 0 ns;
    ciphertext <=
      des_core(plaintext, key_reduce(key), encrypt);
    done <= '1';
  end process;
end;
```

This process is a direct implementation of the main DES routine.
The only implementation specific feature is that the model waits

for the signal go to be raised before starting processing and it raises the signal done at the end of processing, implementing a basic handshaking protocol.

This algorithm requires the two functions: key_reduce and des_core. The former strips the parity bits from the key and the latter then implements the whole DES algorithm. The key_reduce function reduces the key from 64 to 56 bits and permutes the bits to form the initial state of the subkey:

```
function key_reduce(key : in vec64) return vec56 is
--moods inline
begin
  return
  key(57) & key(49) & key(41) & key(33) &
  ...
  key(28) & key(20) & key(12) & key(4);
end;
```

The compiler directive --moods inline causes the synthesizer to inline the function. This allows the optimizer more scope for optimization of the circuit. The des_core function applies the basic DES algorithm 16 times on a slice of the data using a different subkey on each iteration:

```
function des_core
  --moods inline
  (plaintext : vec64;
  key : vec56;
  encrypt : std_logic)
return vec64
is
    variable data : vec64;
    variable working_key : vec56 := key;
begin
  data := initial_permutation(plaintext);
  for round in 0 to 15 loop
    working_key :=
      key_rotate(working_key,round,encrypt);
    data := data(33 to 64) &
      (f(data(33 to 64),key_compress(working_key))
    xor
    data(1 to 32));
  end loop;
  return
    final_permutation(data(33 to 64) & data(1 to 32));
end;
```

The DES algorithm is made up of the key transformation functions key_rotate and key_compress, and the data transformation functions initial_permutation, f and final_permutation.

Data transformations

The data transformations `initial_permutation` and `final_permutation` are simply hard-wired bit-swapping routines implemented using concatenation:

```
function initial_permutation(data : vec64) return vec64 is
  --moods inline
begin
  return
    data(58) & data(50) & data(42) & data(34) &
    ...
    data(31) & data(23) & data(15) & data(7);
end;

function final_permutation(data : in vec64) return vec64 is
  --moods inline
begin
  return
    data(40) & data(8) & data(48) & data(16) &
    ...
    data(49) & data(17) & data(57) & data(25);
end;
```

The `f` function is the main data transform which is applied 16 times to the rightmost half, a 32-bit slice, of the data path. It takes as its second argument a 48-bit subkey generated by the `key_compress` function:

```
function f(data : vec32; subkey : vec48) return vec32 is
  --moods inline
begin
  return permute(substitute(expand(data) xor
  subkey));
end;
```

The function first takes the 32-bit slice of the data path and expands it into 48 bits using the `expand` function. The `expand` function is again just a rearrangement of bits, input bits are replicated in a special pattern to expand the 32-bit input to the 48-bit output:

```
function expand(data : vec32) return vec48 is
  --moods inline
begin
  return
    data(32) & data(1) & data(2) &
    ...
    data(31) & data(32) & data(1);
end;
```

This expanded word is then exclusive-ored with the subkey and fed into a substitute block. This substitutes a different 4-bit pattern for each 6-bit slice of the input pattern (remember that the original input has been expanded from 32 to 48 bits, so there are eight substitutions in all). The substitution also has the effect of reducing the output back to 32 bits again. The substitute algorithm first splits the input 48 bits into eight 6-bit slices. Each slice is then used to look up a substitution pattern for that 6-bit input. This structure is known as the S-block. In the initial implementation, a single Read Only Memory (ROM) is used to store all the substitution patterns. The substitution combines a block index with the input data to form an address which is then used to look up the substitution value in the S-block ROM. This address calculation is encapsulated in the smap function:

```
function smap(index : vec3; data : vec6) return vec4 is
  --moods inline
  type S_block_type is
    array(0 to 511) of natural range 0 to 15;
  constant S_block : S_block_type :=
    --moods ROM
  (
  14, 4, 13, 1, 2, 15, 11, 8, 3, 10, 6, 12, 5, 9, 0, 7,
  ...
  2, 1, 14, 7, 4, 10, 8, 13, 15, 12, 9, 0, 3, 5, 6, 11
  );
begin
  return
    vec4(to_unsigned(S_block(to_integer(unsigned(
index & data(1) & data(6) & data(2 to 5)))), 4));
end;
```

The eight substitutions are carried out by the eight calls to smap in the substitute function:

```
function substitute(data : vec48) return vec32 is
  --moods inline
begin
  return
    smap("000",data(1 to 6)) &
    ...
    smap("111",data(43 to 48));
end;
```

The final stage of the data path transform is the permute function which is another bit-swapping routine:

```
function permute (data : in vec32) return vec32 is
  --moods inline
```

```
begin
  return
    data(16) & data(7) & data(20) & data(21) &
    ...
    data(22) & data(11) & data(4) & data(25);
end;
```

These functions define the whole of the data path part of the algorithm.

Key transformations

The encryption key also needs to be transformed a number of times – specifically, before each data transformation, the key is rotated and then a smaller subkey is extracted by selecting 48 of the 56 bits of the key. The rotation is the most complicated part of the key transformation. The 56-bit key is split into two halves and each half rotated by 0, 1 or 2 bits depending on which round of the DES algorithm is being implemented. The direction of the rotation is to the left during encryption and to the right during decryption. The algorithm is split into two functions: do_rotate which, as the name suggests, does the rotation and key_rotate which calls do_rotate twice, once for each half of the key. The do_rotate function uses a ROM to store the rotate distances for each round, numbered from 0 to 15:

```
function do_rotate
  --moods inline
  (key : in vec28;
  round : natural range 0 to 15;
  encrypt : std_logic)
return vec28 is
  type distance_type is
  array (natural range 0 to 15) of integer range 0 to 2;
  constant encrypt_shift_distance : distance_type :=
  --moods ROM
  (1, 1, 2, 2, 2, 2, 2, 2, 1, 2, 2, 2, 2, 2, 2, 1);
  constant decrypt_shift_distance : distance_type :=
  --moods ROM
  (0, 1, 2, 2, 2, 2, 2, 2, 1, 2, 2, 2, 2, 2, 2, 1);
  variable result : vec28;
begin
  if encrypt = '1' then
  result :=
  vec28(unsigned(key) rol
  encrypt_shift_distance(round));
  else
  result :=
  vec28(unsigned(key) ror
```

```
        decrypt_shift_distance(round));
      end if;
      return result;
    end;
```

The `key_rotate` function simply calls the above function twice:

```
function key_rotate
  --moods inline
  (key : in vec56;
  round : natural range 0 to 15;
  encrypt : std_logic)
  return vec56 is
begin
  return do_rotate(key(1 to 28),round,encrypt) &
    do_rotate(key(29 to 56),round,encrypt);
end;
```

Finally, the key compression function `key_compress` selects 48 of the 56 bits to pass to the S-block algorithm:

```
function key_compress(key : in vec56) return vec48 is
  --moods inline
begin
return
key(14) & key(17) & key(11) & key(24) &
...
key(50) & key(36) & key(29) & key(32);
end;
```

Initial synthesis

The design was synthesized by MOODS with delay prioritized first and area prioritized second. The target technology was the Xilinx Virtex library. Figure 52 shows the control state machine of the synthesized design. The whole state sequence represents the process, which is a loop as shown by the state transition from the last state (c11) back to the first (c1).

The first two states c1 and c2 implement the input handshake on signal go to trigger the process. The DES core is implemented by the remaining states, namely states c3 to c11, which are in the main loop as shown by the state transition back from c11 to c3, so are executed 16 times. There are nine states in this inner loop, giving a total algorithm length of 146 cycles including the two states required for the input handshake and 144 for the DES core itself. However, an inspection of the original structure shown in Figure 51 suggests that a reasonable target for the inner loop is two cycles

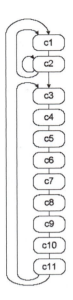

Figure 52
Control State Machine
for Initial Synthesis

per round with an optimistic target of one cycle. Clearly, there is a problem with this design. MOODS predicts that this design has the area and delay characteristics shown in the first table in this chapter in the line labeled (1).

Optimizing the data path

Examining the nine control states in the main loop and relating these to the mapping of the control graph to the DFG showed that the last eight cycles were performing the S-block and the first two cycles were mainly related to transforming the key. The second state is an overlap state where both key and data transforms are taking place. The problem with the last eight cycles was fairly self-evident since there are eight substitutions and there are eight control states to perform them. Clearly, there was something causing each substitution to be locked into a separate control state and therefore preventing optimization with respect to latency. It wasn't difficult to see what – each of these states contained just register assignments, concatenations and a ROM read operation. It is the last of these that is the problem – the ROM implementation being targeted is a synchronous circuit, so the S-block ROM can only be accessed once per clock cycle – in other words once per control state. It is this that is preventing the data path operations from being performed in parallel.

Attacking this problem is beyond the capabilities of behavioral synthesis because it requires knowledge of the dataflow at a much higher level than can be automatically extracted. The solution therefore requires modification of the original design.

There are two obvious solutions to this problem – either split the S-block into eight smaller ROMs that can therefore be accessed in parallel or make the S-block a non-ROM so that the array gets expanded into a decoder block once for each access, giving eight decoders. The latter solution appears simplest, but it will result in eight 512-way decoders, which will be a very large implementation. The solution of splitting the ROMs is more likely to yield a useful solution. The substitute function was rewritten to have eight mini-ROMs:

```
function substitute(data : vec48) return vec32 is
  --moods inline
  type S_block_type is
    array(0 to 63) of natural range 0 to 15;
  constant S_block0 : S_block_type := ( ... );
  --moods ROM
  ...
  constant S_block7 : S_block_type := ( ... );
  --moods ROM
begin
  return std_logic_vector(to_unsigned(S_block0(to_integer(
    unsigned(data(1) & data(6) & data(2 to 5)))),4)) &
    ...
    std_logic_vector(to_unsigned(S_block7(to_integer(
    unsigned(data(43) & data(48) & data(44 to 47)))),4));
end;
```

This was resynthesized and resulted in the control graph shown in Figure 53. The inner loop was found to have been reduced to two states, and examination of the last state confirmed that all of the S-block substitutions were being carried out in the one state

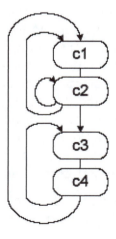

Figure 53
Control State
Machine for
Optimized S-blocks

c4. The key transformations were still split across the two inner states c3 and c4.

One interesting side effect of this optimization is that it is also a smaller design. MOODS predicts that this design has the area and delay characteristics shown in the table (section Results) in the line labeled (2).

Optimizing the key transformations

Examination of the two control states in the main loop, which both contain key transformations, showed that both of these states were performing ROM access and rotate operations. Examination of the original `key_rotate` function showed that the shift distance ROMs are accessed twice per call, so this turned out to be exactly the same problem as with the S-block ROM. Since ROMs are synchronous, they can only be accessed once per cycle and this forces at least two cycles to be used for the rotate. To solve this, the function can be rewritten to only access the ROMs once per call:

```
if encrypt = '1' then
  distance := encrypt_shift_distance(round);
  result :=
    vec28(unsigned(key(1 to 28)) rol distance) &
    vec28(unsigned(key(29 to 56)) rol distance);
else
  distance := decrypt_shift_distance(round);
  result :=
    vec28(unsigned(key(1 to 28)) ror distance) &
    vec28(unsigned(key(29 to 56)) ror distance);
end if;
```

This was resynthesized and resulted in the control graph shown in Figure 54. The inner loop was found to have been reduced to one state (c3) containing both the key and data transformations

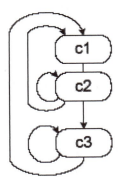

Figure 54
Control State Machine for Optimized Key Rotate

which are repeated 16 times. As before, states c1 and c2 implement the input handshake.

So, this optimization means that the target of 1 clock cycle per round of the core was achieved. MOODS predicts that this design has the area and delay characteristics shown in the table (section Results) in the line labeled (3).

Final optimization

It was recognized that the `key_rotate` function could be simplified by rethinking the rotate algorithm such that a right rotate of 1 bit was replaced by a left rotate of 27 bits (for a 28-bit word). This eliminates a conditional statement, which it was felt could be preventing some optimizations from taking place. This means that there was no need to have a different algorithm for encryption and decryption. This led to the following rework:

```
function key_rotate
  --moods inline
  (key : vec56;
  round : natural range 0 to 15;
  encrypt : std_logic)
return vec56 is
  type distance_type is
  array (natural range 0 to 31) of integer range
  0 to 31;
constant shift_distance : distance_type :=
  --moods ROM
  (0, 1, 2, 2, 2, 2, 2, 2,
  1, 2, 2, 2, 2, 2, 2, 1,
  27, 27, 26, 26, 26, 26, 26, 26,
  27, 26, 26, 26, 26, 26, 26, 27);
  variable distance : natural range 0 to 31;
begin
  distance := shift_distance(to_integer(unsigned(
    encrypt & to_unsigned(round,4))));
  return vec28(unsigned(key(1 to 28)) ror distance) &
    vec28(unsigned(key(29 to 56)) ror distance);
end;
```

The state machine for this design was basically the same as for the previous design as shown in Figure 54. It was found that this version was slightly slower than the previous design but significantly smaller.

MOODS predicts that this design has the area and delay characteristics shown in the table (section Results) in the line labeled (4).

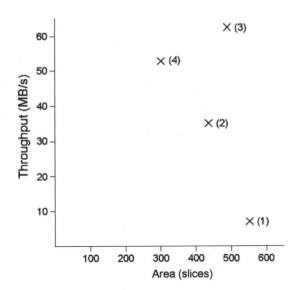

Figure 55
Area vs. Throughput
for All DES Designs

Results

The results predicted by MOODS for all the variations of the design discussed so far are summarized in the table below:

Physical Metrics for Single DES Designs				
Design	**Area (slices)**	**Latency (cycles)**	**Clock (ns)**	**Throughput (MB/s)**
(1) Initial Design	552	146	7.8	7.12
(2) Optimized S-Blocks	426	34	7.1	35.2
(3) Optimized Key	489	18	7.1	62.6
(4) Optimized Branch	307	18	8.4	52.9

It can be seen that design (3) is the fastest, but design (4) is the smallest. Figure 55 plots area vs. throughput for these 4 designs. The *X*-axis represents the area of the design and the *Y*-axis the throughput.

Triple DES

Introduction

Building on this, the DES core developed above was used as the core for a triple DES implementation.

The idea of triple DES is that data is encrypted 3 times. The rationale for choosing three iterations and the advantages and the disadvantages of this are explained by Scheier in *Applied Cryptography*. A common form of triple DES is known as EDE2, which means data is encrypted, decrypted and then encrypted again using two different keys. The first key is used for both encryptions and the second key for the decryption. There are obviously a number of different trade-offs that can be made in this design. Each of these is examined in the following sections. In all cases, the smallest implementation (design (4)) was used as the DES core.

Minimum area: iterative

To achieve a minimum area implementation, a single DES core is used for all three stages. The data is passed through this core 3 times with the different permutations of keys and encryption mode to achieve the EDE2 algorithm. Two different styles of VHDL were tried. These differed in the method used to select the different inputs for each encryption step. The first style used a case statement and the second style used indexed arrays. The case statement style results in the following VHDL design:

```vhdl
library ieee;
use ieee.std_logic_1164.all;
entity tdes_ede2_iterative is
  port(
    plaintext : in std_logic_vector(1 to 64);
    key1 : in std_logic_vector(1 to 64);
    key2 : in std_logic_vector(1 to 64);
    encrypt : in std_logic;
    go : in std_logic;
    ciphertext : out std_logic_vector(1 to 64);
    done : out std_logic);
end;
architecture behavior of tdes_ede2_iterative is
  ...
begin
  process
    variable data : vec64;
    variable key : vec56;
    variable mode : std_logic;
  begin
    wait until go = '1';
    done <= '0';
    wait for 0 ns;
    data := plaintext;
  for i in 0 to 2 loop
    case i is
```

```
            when 1=>
              key := key_reduce(key2);
              mode := not encrypt;
            when others =>
              key := key_reduce(key1);
              mode := encrypt;
            end case;
            data := des_core(data,key,mode);
          end loop;
          ciphertext <= data;
          done <= '1';
        end process;
      end;
```

It can be seen that this uses a case statement to select the appropriate key and encryption mode for each iteration. The characteristics of the case statement solution are shown in the table (section Comparing the Approaches) in the line labeled (5).

The core DES algorithm accounts for 48 cycles (3 iterations of 16 rounds with 1 cycle per round), leaving an additional overhead of 3 cycles. This additional 3 cycles is due to the case statement selection of the key which adds an extra cycle per iteration of the core. The second style uses arrays to store the keys and modes and then indexes these arrays to set the key and mode for each iteration. The process becomes:

```
process
  ...
  type keys_type is array (0 to 2) of vec56;
  variable keys : keys_type;
  type modes_type is array (0 to 2) of std_logic;
  variable modes : modes_type;
begin
  ...
  modes := (encrypt, not encrypt, encrypt);
  keys := (key_reduce(key1),
    key_reduce(key2),
    key_reduce(key1));
  for i in 0 to 2 loop
    data := des_core(data,keys(i),modes(i));
  end loop;
  ...
```

It was found that the latency was the same as the case statement solution but the area was approximately 25 per cent larger. This overhead is mostly due to the use of the register arrays which add up to about 200 extra flip-flops. Clearly the case statement design is the most efficient of the two and so this solution was kept and the array style solution discarded.

Minimum latency: pipelined

To achieve minimum latency between samples, three DES cores
are used to form a pipeline. Data samples can then be fed into the
pipeline every 18 cycles (the latency of the single core), although
the time taken for a result to be generated is 50 cycles because of
the pipeline length. The circuit is simply three copies of the single
DES process:

```
architecture behavior of tdes_ede2_pipe is
  ...
  signal intermediate1, intermediate2 : vec64;
begin
  process
  begin
    wait until go = '1';
    intermediate1 <=
      des_core(plaintext,key_reduce(key1),encrypt);
  end process;
  process
  begin
    wait until go = '1';
    intermediate2 <=
      des_core(intermediate1,key_reduce(key2),not
      encrypt);
  end process;
  process
  begin
    wait until go = '1';
    done <= '0';
    wait for 0 ns;
    ciphertext <=
      des_core(intermediate2,key_reduce(key1),
      encrypt);
    done <= '1';
  end process;
end;
```

Note how the done output is driven only by one of the cores – this
will give the right result provided all three cores synthesize to the
same delay, which in practice they will. This design decision alle-
viates the need to have handshaking between the cores. MOODS
predicts that this design has the area and delay characteristics
shown in the table (section Comparing the Approaches) in the line
labeled (6). The state machine in Figure 56 shows the three independent processes. For example, the first process is represented by
states c2, c3 and c4. The first two states perform the handshaking
on go and c4 implements the DES core with its 16 iterations. State
c7 is the second DES core and c10 the third.

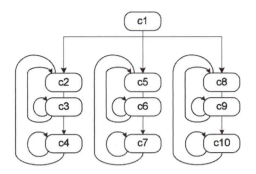

Figure 56
Control State
Machine for Pipelined
Triple DES

Comparing the approaches

The physical metrics of the previous section are the predicted values given by MOODS. To get a more accurate assessment of the

Design	Tool	Area (slices)	Latency (cycles)	Clock (ns)	Throughput (MB/s)
DES	MOODS	307	18	8.4	52.9
	Leonardo	258		13.4	33.2
	Foundation	274		18.4	24.2
Iterative TDES	MOODS	500	53	8.4	18.0
	Leonardo	381		13.7	11.0
	Foundation	422		17.8	8.5
Pipelined TIDES	MOODS	920	18	8.4	52.9
	Leonardo	774		13.7	32.4
	Foundation	826		18.4	24.2

design, RTL synthesis of the structural VHDL output of MOODS is required. This was carried out using Mentor Graphics' Leonardo Spectrum RTL synthesis suite. These results can be finessed further by carrying out placement and routing using the Xilinx Integrated Software Environment (ISE) Foundation suite. The results predicted by all three tools (MOODS, Leonardo and Foundation) for the three approaches (DES, Iterative TDES and Pipelined TDES) are shown in the table below. In all cases, the design was optimized during RTL synthesis using the vendor's default optimization settings – a combination of area and delay optimization – with maximum optimization effort. Placement and routing was performed with an unreachable clock period to force Foundation to produce the fastest design.

This shows that MOODS tends to overestimate the area of the design and underestimate the delay. Both of these are expected outcomes. The tendency to overestimate area is because it isn't possible to predict the effect of logic minimization when working at the behavioral level. The tendency to underestimate delay is because it isn't possible to predict routing delays.

Summary

This chapter has shown that it is possible to design complex algorithms such as DES using the abstraction of high-level VHDL and get a synthesizable design. However, the synthesis process is not and cannot ever be fully automated – human guidance is still necessary to optimize the design's structure to get the best from the synthesis tools. Nevertheless the modifications are high-level design decisions and the final design is still readable and abstract. There has been no need to descend to low-level VHDL to implement DES. The implementations of triple DES show how VHDL code can easily be reused when written at this level of abstraction. It is quite an achievement to implement the DES and two implementations of the triple DES algorithm in 4 working days including testing and this demonstrates the kind of productivity that result from the application of behavioral synthesis tools.

Part 5
Fundamental Techniques

In this fifth part of the book, the aim is to present a collection of standard functions in VHDL that are quite basic. This is directed to those who perhaps are new to VHDL and need even simple functions to be provided for them in VHDL. This part of the book describes standard techniques for implementing registers, counters, decoders, multiplexers, latches and flip flops, and also covers background information such as fixed point arithmetic operations, binary multiplication, finite state machines, serial to parallel and parallel to serial conversion and ALU functions.

The VHDL provided is 'examplar' in that clarity and simplicity were preferred over efficiency, speed or area and as such a practical implementation will require optimisation and further design. The VHDL is designed to enable a designer to *understand* how these operation work and to implement their own functions in the light of that knowledge.

19
Counters

Introduction

One of the most commonly used applications of flip-flops in practical systems is counters. They are used in microprocessors to count the program instructions (program counter or PC), for accessing sequential addresses in memory (such as ROM) or for checking progress of a test. Counters can start at any value, although most often they start at zero and they can increment or decrement. Counters may also change values by more than one at a time, or in different sequences (such as gray code, binary or Binary Coded Decimal (BCD) counters).

Basic binary counter

The simplest counter to implement in many cases is the basic binary counter. The basic structure of a counter is a series of flip-flops (a register), that is controlled by a reset (to reset the counter to zero literally) and a clock signal used to increment the counter. The final signal is the counter output, the size of which is determined by the generic parameter n, which defines the size of the counter. The symbol for the counter is given in Figure 57. Notice that the reset is active low and the counter and reset inputs are given in a separate block of the symbol as defined in the IEEE standard format.

From an Field Programmable Gate Array (FPGA) implementation point of view, the value of generic *n* also defines the number of D type flip-flops required (usually a single LUT) and hence the

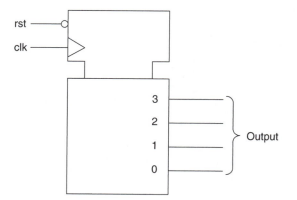

Figure 57
Simple Binary
Counter

usage of the resources in the FPGA. A simple implementation of such a counter is given below:

```
library ieee;
use ieee.std_logic_1164.all;
use ieee.numeric_std.all;

entity counter is
  generic (
       n : integer := 4);
  port (
       clk : in std_logic;
       rst : in std_logic;
       output : out std_logic_vector((n-1) downto 0)
  );
end;


architecture simple of counter is
begin
  process(clk, rst)
   variable count : unsigned((n-1) downto 0);
  begin
   if rst= '0' then
     count := (others => '0');
   elsif rising_edge(clk) then
     count := count + 1;
   end if;
   output <= std_logic_vector(count);
  end process;
end;
```

The important aspect of this approach to the counter VHDL is that this is effectively a state machine, however we do not have to explicitly define the next state logic – this will be taken care of by the synthesis software. This counter can now be tested using a simple test bench that resets the counter and then clocks the state machine until the counter flips round to the next counter round. The test bench is given below:

```vhdl
library ieee;
use ieee.std_logic_1164.all;
use ieee.numeric_std.all;

entity CounterTest is
end CounterTest;

architecture stimulus of CounterTest is
      signal rst : std_logic := '0';
      signal clk : std_logic := '0';
      signal count : std_logic_vector (3 downto 0);

      component counter
          port(
                clk : in std_logic;
                rst : in std_logic;
                output : out std_logic_vector(3 downto 0)
          );
    end component;
    for all : counter use entity work.counter ;

begin
    CUT: counter port map(clk=>clk,rst=>rst,
      output=>count);
    clk <= not clk after 1 us;
    process
    begin
    rst <= '0','1' after 2.5 us;
    wait;
    end process;
end;
```

Using this simple VHDL testbench, we reset the counter until 2.5 us and then the counter will count on the rising edge of the clock after 2 us (i.e. the counter is running at 500 kHz).

If we dissect this model, there are several interesting features to notice. The first is that we need to define an internal variable **count** rather than simply increment the output variable q. The output variable q has been defined as a standard logic vector (std_logic_vector) and with it being defined as an output we cannot use it as the input variable to an equation. Therefore we need to define a local variable (in this case count) to store the current value of the counter.

The initial decision to make is should we use a variable or a signal? In this case, we need an internal variable that we can effectively treat as a sequential signal, and also one that changes instantaneously, which immediately requires the use of a variable. If we chose a signal, then the changes would only take place when the cycle is resolved (i.e. the next time the process is activated).

The second decision is what type of unit should the count variable be? The output variable is a std_logic_vector type which has the advantage of being an array of std_logic signals, and so we don't need to specify the individual bits on a word, this is done automatically. The major disadvantage, however, is that the std_logic_vector does not support simple arithmetic operations, such as addition, directly. In this example, we want the counter to have a simple definition in VHDL and so the best compromise type that has the bitwise definition and also the arithmetic functionality would be the unsigned or signed type. In this case, we wish to have a direct mapping between the std_logic_vector bits and the counter bits, so the unsigned type is appropriate. Thus the declaration of the internal counter variable count is as follows:

```
variable count : unsigned((n-1) downto 0);
```

The final stage of the model is to assign the internal value of the count variable to the external std_logic_vector q. Luckily, the translation from unsigned to std_logic_vector is fairly direct, using the standard casting technique:

```
q <= std_logic_vector(count);
```

As the basic types of both q and count are consistent, this can be done directly.

Synthesized simple binary counter

At this point it is useful to consider what happens when we synthesize this VHDL, so to test this point the VHDL model of the simple binary counter was run through a typical RTL synthesis software package (Leonardo Spectrum) with the resultant synthesized VHDL model given below:

```
entity counter is
    port (
        clk : IN std_logic;
        rst : IN std_logic;
        output : OUT std_logic_vector (3 DOWNTO 0));
end counter;
```

```
architecture simple of counter is
    signal clk_int, rst_int, output_dup0_3,
      output_dup0_2, output_dup0_1,
      output_dup0_0, output_nx4, output_nx7,
      output_nx10, NOT_rst,
      output_NOT_a_0: std_logic;

  begin
    output_obuf_0 : OBUF port map ( O=>output(0),
      I=>output_dup0_0);
    output_obuf_1 : OBUF port map ( O=>output(1),
      I=>output_dup0_1);
    output_obuf_2 : OBUF port map ( O=>output(2),
      I=>output_dup0_2);
    output_obuf_3 : OBUF port map ( O=>output(3),
      I=>output_dup0_3);
    rst_ibuf : IBUF port map ( O=>rst_int, I=>rst);
    output_3_EXMPLR_EXMPLR : FDC port map
      (Q=>output_dup0_3, D=>output_nx4,
      C=>clk_int, CLR=>NOT_rst);
    output_2_EXMPLR_EXMPLR : FDC port map
      (Q=>output_dup0_2, D=>output_nx7,
      C=>clk_int, CLR=>NOT_rst);
    output_1_EXMPLR_EXMPLR : FDC port map
      (Q=>output_dup0_1, D=>output_nx10,
      C=>clk_int, CLR=>NOT_rst);
    output_0_EXMPLR_EXMPLR : FDC port map
      (Q=>output_dup0_0, D=> output_NOT_a_0,
      C=>clk_int, CLR=>NOT_rst);
    clk_ibuf : BUFGP port map ( O=>clk_int, I=>clk);
    output_nx4 <= (not output_dup0_3 and output_dup0_2
      and output_dup0_1 and output_dup0_0) or
      (output_dup0_3 and not output_dup0_0) or
      (output_dup0_3 and not output_dup0_2) or
      (output_dup0_3 and not output_dup0_1);
    output_nx7 <= (output_dup0_2 and not output_dup0_0)
      or (not output_dup0_2 and output_dup0_1 and
      output_dup0_0) or (output_dup0_2 and not
      output_dup0_1);
    output_nx10 <= (output_dup0_0 and not
      output_dup0_1) or (not output_dup0_0 and
      output_dup0_1);
    NOT_rst <= (not rst_int);
    output_NOT_a_0 <= (not output_dup0_0);
  end simple;
```

The first obvious aspect of the model is that it is much longer than the simple RTL VHDL created originally. The next stage logic is now in evidence, as this is synthesized, the physical gates must be defined for the model. Finally the outputs are buffered which leads to even more gates in the final model. If the optimization report is observed, the overall statistics of the resource usage of the FPGA can be examined (in this case a Xilinx Virtex-II Pro device):

```
Cell    Library References       Total Area

===================================================

BUFGP   xcv2p    1 x         1   1 BUFGP
FDC     xcv2p    4 x         1   4 Dffs or Latches
IBUF    xcv2p    1 x         1   1 IBUF
LUT1    xcv2p    2 x         1   2 Function Generators
LUT2    xcv2p    1 x         1   1 Function Generators
LUT3    xcv2p    1 x         1   1 Function Generators
LUT4    xcv2p    1 x         1   1 Function Generators
OBUF    xcv2p    4 x         1   4 OBUF

Number of ports :                      6
  Number of nets :                              17
  Number of instances :                         15
  Number of references to this view :            0
Total accumulated area :
  Number of BUFGP :                              1
  Number of Dffs or Latches :                    4
  Number of Function Generators :                5
  Number of IBUF :                               1
  Number of OBUF :                               4
  Number of gates :                              5
  Number of accumulated instances :             15
Number of global buffers used:                   1
************************************************
Device Utilization for 2VP2fg256
************************************************
Resource               Used    Avail   Utilization
--------------------------------------------------
IOs                     5       140     3.57%
Global Buffers          1        16     6.25%
Function Generators     5      2816     0.18%
CLB Slices              3      1408     0.21%
Dffs or Latches         4      3236     0.12%
Block RAMs              0        12     0.00%
Block Multipliers       0        12     0.00%
```

In this simple example, it can be seen that the overall utilization of the FPGA is minimal, with the relative resource allocation of IOs, buffers and functional blocks. This is an important aspect of FPGA design in that even though the overall device may be underutilized, a particular resource (such as IO) might be used up. The output VHDL can then be used in a physical place and route software tool (such as the Xilinx Design Navigator) to produce the final bit file that will be downloaded to the device.

Shift register

While a shift register is strictly speaking not a counter, it is useful to consider this in the context of other counters as it can be converted into a counter with very small changes. We will consider this element layer in this book, in more detail, but consider a simple case of a shift register that takes a single bit and stores in the least significant bit of a register and shifts each bit up one bit on the occurrence of a clock edge. If we consider an n-bit register and show the status before and after a clock edge, then the functionality of the shift register becomes clear as shown in Figure 58.

A basic shift register can be implemented in VHDL as shown below:

```
library ieee;
use ieee.std_logic_1164.all;

entity shift_register is
  generic (
      n : integer := 4);
  port (
        clk : in std_logic;
        rst : in std_logic;
        din : in std_logic;
        q : out std_logic_vector((n-1) downto 0)
  );
end entity;
```

(a)

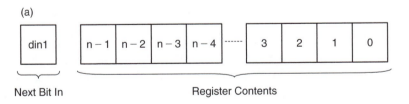

(b)

Figure 58
Shift Register Functionality: (a) before and (b) after the clock edge

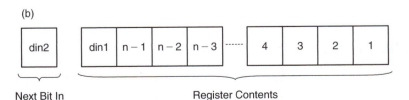

```
architecture simple of shift_register is
begin
  process(clk, rst)
    variable shift_reg : std_logic_vector((n-1) downto 0);
  begin
    if rst= '0' then
      shift_reg := (others => '0');
    elsif rising_edge(clk) then
      shift_reg := shift_reg(n-2 downto 0) & din;
    end if;
    q <= shift_reg;
  end process;
end architecture simple;
```

The interesting parts of this model are very similar to the simple binary counter, but subtly different. As for the counter, we have defined an internal variable (shift_reg), but unlike the counter we do not need to carry out arithmetic functions, so we do not need to define this as an unsigned variable, but instead we can define directly as a std_logic_vector – the same as the output q.

Notice that we have an asynchronous clock in this case. As we have discussed previously in this book, there are techniques for completely synchronous sets or resets, and these can be used if required.

The fundamental difference between the counter and the shift register is in how we move the bits around. In the counter we use arithmetic to add one to the internal counter variable (count). In this case we just require to shift the register up by one bit, and to achieve this we simply assign the lowest $(n-1)$ bits of the internal register variable (shift_reg) to the upper $(n-1)$ bits and concatenate the input bit (din), effectively setting the lowest bit of the register to the input signal (din). This is accomplished using the VHDL below:

```
shift_reg := shift_reg(n-2 downto 0) & din;
```

The final stage of the model is similar to the basic counter in that we then assign the output signal to the value of the internal variable (shift_reg) using a standard signal assignment. In the shift register, we do not need to 'cast' the type as both the internal and signal variable types are std_logic_vector:

```
q <= shift_reg;
```

The Johnson counter

The Johnson counter is a counter that is a simple extension of the shift register. The only difference between the two is that the

Johnson counter has its least significant bit inverted and fed back into the most significant bit of the register. In contrast to the classical binary counter that has 2^n states, the Johnson counter has 2^n states. While this has some specific advantages, a disadvantage is that the Johnson counter has what is called a 'parasitic counter' in the design. In other words, while the 2^n counter is operating, there is another state machine that also operates concurrently with the Johnson counter using the unused states of the binary counter.

A potential problem with this counter is that if, due to an error, noise or other glitch, the counter enters a state NOT in the standard Johnson counting sequence, it cannot return to the correct Johnson counter without a reset function. The normal Johnson counter sequence is shown in the following table:

Count	Q(3:0)
0	0000
1	1000
2	1100
3	1110
4	1111
5	0111
6	0011
7	0001

The VHDL implementation of a simple Johnson counter can then be made by modifying the next stage logic of the internal shift_register function as shown below:

```
library ieee;
use ieee.std_logic_1164.all;

entity johnson_counter is
  generic (
       n : integer := 4);
  port (
       clk : in std_logic;
       rst : in std_logic;
       din : in std_logic;
       q : out std_logic_vector((n-1) downto 0)
  );
end entity;
```

```
architecture simple of Johnson_counter is
begin
  process(clk, rst)
    variable j_state : std_logic_vector((n-1) downto 0);
  begin
    if rst= '0' then
      j_state:= (others => '0');
    elsif rising_edge(clk) then
      j_state:= not j_state(0) & j_state(n-1 downto 1);
    end if;
    q <= j_state;
  end process;
end architecture simple;
```

Notice that the concatenation is now putting the inverse (NOT) of the least significant bit of the internal state variable (j_state(0)) onto the next state most significant bit, and then shifting the current state *down* by one bit.

It is also worth noting that the counter does not have any checking for the case of an incorrect state. It would be sensible in a practical design to perhaps include a check for an invalid state and then reset the counter in the event of that occurrence. The worst case scenario would be that the counter would be incorrect for a further 7 clock cycles before correctly resuming the Johnson counter sequence.

BCD counter

The BCD counter is simply a counter that resets when the decimal value 10 is reached instead of the normal 15 for a 4 bit binary counter. This counter is often used for decimal displays and other human interface hardware. The VHDL for a BCD counter is very similar to that of a basic binary counter except that the maximum value is 10 (hexadecimal A) instead of 15 (hexadecimal F). The VHDL for a simple BCD counter is given below. The only change is that the counter has an extra check to reset when the value of the count variable is greater than 9 (the counter range is 0 to 9).

```
library ieee;
use ieee.std_logic_1164.all;
use ieee.numeric_std.all;

entity counter is
  generic (
      n : integer := 4);
```

```
      port (
            clk : in std_logic;
            rst : in std_logic;
            output : out std_logic_vector((n-1) downto 0)
      );
end;

architecture simple of counter is
begin
  process(clk, rst)
  variable count : unsigned((n-1) downto 0);
  begin
  if rst= '0' then
    count := (others => '0');
  elsif rising_edge(clk) then
    count := count + 1;
    if count > 9 then
          count := 0;
    else if
  end if;
  output <= std_logic_vector(count);
 end process;
end;
```

Summary

In this chapter, we have investigated some basic counters and shown how VHDL can be used to carry out arithmetic functions or logic functions to obtain the required counting sequence. The possibilities of counters based on these basic types are numerous, possibly infinite, and it is left to the reader to develop their own variations based on these standard types.

A useful exercise would be to modify the basic binary counter by adding an up/down flag so that depending on this flag, the counter would increment or decrement, respectively.

20

Latches, Flip-Flops and Registers

Introduction

There are different types of storage elements that will occur from different VHDL code, and it is important to understand each of them, so that the correct one results when a design is synthesized. Often bugs in hardware happen due to misunderstanding about what effect a VHDL construct will have on the resulting synthesized hardware. In this chapter we will introduce three main types of storage elements used in VHDL that can be synthesized from VHDL to an Field Programmable Gate Array (FPGA) platform, latches, flip-flops and registers.

Latches

A latch can be simply defined as a level sensitive memory device. In other words, the output depends purely on the value of the inputs. There are several different types of latch, the most common being the D Latch and the SR Latch.

First consider a simple D latch as shown in Figure 59. In this type of latch, the output (Q) follows the input (D), but only when the Enable (En) is high. In this case we are referring to a D latch, and the full definition is in fact a level sensitive D latch. The assumption

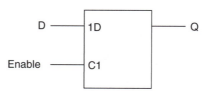

Figure 59
D Latch Symbol

made in this book is that whenever we refer to a latch is that it is always level sensitive.

The notation on the Enable signal (C1) and the Data input (1D) denote that they are linked together. Also notice that the output Q is purely dependent on the level of D and the Enable. In other words, when the Enable is high, then Q = D. This is called a level sensitive latch.

The VHDL that represents this kind of level sensitive D latch is shown below:

```
library ieee;
use ieee.std_logic_1164.all;
entity latch is
  port ( d :   in std_logic;
         en :  in std_logic;
         q :   out std_logic);
end entity latch;

architecture beh of latch is
begin
    process (d, en) is
    begin
        if (en = '1') then
        q <= d;
        end if;
    end process;
end architecture beh;
```

This is an example of an 'incomplete if' statement, where the condition 'if (en = '1')' is given, but the 'else' condition is not defined. Both **d** and **en** are in the sensitivity list and so this could be combinatorial, but due to the incomplete definition of **en**, then an implied latch occurs, i.e. storage.

This aspect of storage is important when we are developing models, particularly behavioral as in this case (i.e. the structure is not explicitly defined), as we may end up with latches in our design even though we *think* that we have created a model for a purely combinatorial circuit.

Other instances when this may occur are the incomplete definition of case statements. For example, consider the simple VHDL example below:

```
case s is
  when "00" => y <= a;
  when "10" => y <= b;
  when others => null;
end case;
```

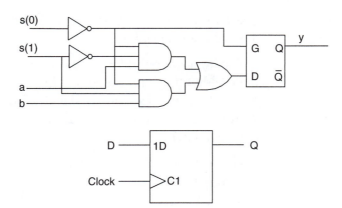

Figure 60
Synthesised Latch

Figure 61
D-Type Flip-Flop

In this statement, it is incomplete and so instead of a simple combinatorial circuit, a latch is implied. The resulting synthesized circuit is shown below in Figure 60.

Flip-flops

In contrast to the level triggered latch, the flip-flop changes state when an edge occurs on an enable or a clock signal. This is the cornerstone of synchronous design, with an important building block being the D-type flip-flop as shown in Figure 61. The output (Q) will take on the value of the input (D) on the rising edge of the clock signal. The triangle on the symbol denotes a clock signal and in the absence of a circle (notation for active low), the definition is for a rising edge to activate the flip-flop.

The equivalent VHDL code is of the form shown below:

```
library ieee;
use ieee.std_logic_1164.all;
entity dff is
  port ( d :     std_logic;
         clk :   in std_logic;
         q :     out std_logic);
end entity dff;

architecture simple of dff is
begin
process (clk) is
  begin
     if rising_edge(clk) then
        q <= d;
     end if;
end process;
end architecture simple;
```

Notice that in this case, d does not appear in the sensitivity list – it is not required. The flip-flop will only do something when a rising edge occurs on the clock signal (clk). There are a number of different methods of describing this functionality, all of them equivalent. In this case, we have explicitly defined the clk signal in the sensitivity list. An alternative method would be to have no sensitivity list, but to add a wait on statement inside the process. The equivalent architecture would be as follows:

```
architecture wait_clk of dff is
begin
     process is
     begin
          if rising_edge(clk) then
                q <= d;
          end if;
          wait on clk;
     end process;
end architecture simple;
```

We have also perhaps used a more complex definition of the rising_edge function than is required (or may be available in all simulators or synthesis tools). The alternative simple method is to use the clock in the sensitivity list and then check that the value of clock is '1' for rising edge or '0' for falling edge. The equivalent VHDL for a rising edge D-type flip-flop is given below. Notice that we have used the implicit sensitivity list (using a wait on clk statement) as opposed to the explicit sensitivity list, although we could use either interchangeably.

```
architecture rising_edge_clk of dff is
begin
  process is
  begin
    if (clk = '1') then
          q <= d;
    end if;
    wait on clk;
  end process;
end architecture simple;
```

We can extend this basic model for a D-type to include an asynchronous set and reset function. If they are asynchronous, this means that they could happen whether there is a clock edge or not, therefore they need to be added to the sensitivity list of the model.

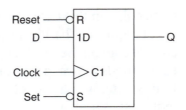

Figure 62
D-Type Flip-Flop
with Asynchronous
Set and Reset

The symbol for such a flip-flop assuming active low set and reset would be as shown in Figure 62.

The VHDL is extended from the simple dff model previously given to include the asynchronous set and reset as shown below:

```
library ieee;
use ieee.std_logic_1164.all;
entity dff_sr is
  port ( d    : in std_logic;
         clk  : in std_logic;
         nrst : std_logic;
         nset : in std_logic;
         q    : out std_logic);
end entity dff_sr;

architecture simple of dff_sr is
begin
process (clk, nrst, nset) is
  begin
    if (nrst = '0') then
      q <= '0';
    elsif (nset = '1') then
      q <= '1';
    elsif rising_edge(clk) then
      q <= d;
    end if;
end process;
end architecture beh;
```

As for the basic D-type flip-flops, we could use a variation of the check for the clock edge, although due to the fact that we have three possible input state control variables (set, reset and clk) it is not enough now to check whether the clock is high (for a rising edge flip-flop). It is necessary to check that the clock is high *and* that an event has occurred.

Notice that this model may cause interesting behavior when synthesized as the reset will *always* be checked before the set and so there is a specific functionality that allows the concurrent setting of the set and reset variables, but the reset will take precedence.

Finally, when considering the use of transitions between '0' and '1', there are issues with synthesis and simulation when using the different approaches. For example, with the standard logic package (std_logic variables), the transitions are strictly defined and so we may have the case of high impedance or don't care states occurring during a transition. This is where the rising_edge (and its opposite the falling_edge) function are useful as they simplify all these options into a single function that handles all the possible transition states cleanly.

It is generally best, therefore, to use the rising_edge or falling_edge functions wherever possible to ensure consistent and interoperable functionality of models.

It is also worth considering a synchronous set or reset function, so that the clock will be the only edge that is considered. The only caveat with this type of approach is that the set and reset signals should be checked immediately following the clock edge to make sure that concurrent edges on the set or reset signals have not occurred.

Registers

Registers use a bank of flip-flops to load and store data in a bus. The difference between a basic flip-flop and a register is that while there is a data input, clock and usually a reset (or clear), there is also a 'load' signal that defines whether the data on the input is to be loaded onto the register or not. The VHDL code for an example 8-bit register would be as follows:

```
library ieee;
use ieee.std_logic_1164.all;
entity register is
  generic (n : natural := 8);
  port (d : in std_logic_vector(n-1 downto 1);
        clk : in std_logic;
        nrst : in std_logic;
        load : in std_logic;
        q : out std_logic_vector(n-1 downto 1));
end entity register;

architecture beh of register is
begin
process (clk, nrst) is
  begin
    if (nrst = '0') then
        q <= (others => '0');
```

```
      elsif (rising_edge(Clock) and (load = '1')) then
         q <= d;
      end if;
   end process;
end architecture beh;
```

Notice that although there are four inputs (clk, nrst, load and d), only clk and nrst are included in the process sensitivity list. If load and d change, then the process will ignore these changes until the clk rising edge or nrst goes low. If the load is not used, then the register will load the data on every clock rising edge unless the reset is low. The VHDL for this slightly simpler register is given below:

```
library ieee;
use ieee.std_logic_1164.all;
entity reg_rst is
  port (d, clk, nrst : in std_logic;
        q : out std_logic);
end entity reg_rst;

architecture beh of reg_rst is
begin
process (clk, nrst) is
  begin
    if (nrst = '0') then
       q <= '0';
    elsif rising_edge(clk) then
       q <= d;
    end if;
end process;
end architecture beh;
```

Summary

In this chapter the basic type of latch and register have been introduced and examples given. This is a fundamental building block of synchronous digital systems and is the basis of RTL (Register Transfer Logic) design with VHDL.

21

Serial to Parallel & Parallel to Serial Conversion

Serial to Parallel Conversion

Serial to Parallel Conversion (SIPO) is a relatively simple matter of clocking in a single bit stream into a register and shifting each bit in turn until the register is full. Then the parallel output can be read directly. In this example VHDL model, the size of the register is set by the generic (n), which in this case defaults to 8. Notice that in this example, the reset signal (nrst) is synchronous, not asynchronous as has been used before. In this case, the only signal that the process will react to is an event on the clock (clk), and a rising_edge event at that. When this event occurs, the reset signal is checked to see if it is low, otherwise the register is clocked through. If the reset signal is low, then the register is cleared to all zeros.

```
LIBRARY ieee;
USE ieee.Std_logic_1164.ALL;
USE ieee.Std_logic_unsigned.ALL;

ENTITY sipo IS
      GENERIC(n : Positive := 8);
      PORT(
                  clk : in std_logic;
                  nrst : in std_logic;
                  di : in std_logic;
                  q: out std_logic_vector((n-1) DOWNTO 0)
          );
END sipo;

ARCHITECTURE simple OF sipo IS
   SIGNAL int_reg : Std_logic_vector((n-1) DOWNTO 0);
   signal index : integer := 0;
```

```
BEGIN
    out_process : PROCESS
BEGIN
        WAIT UNTIL rising_edge(clk);
        if nrst = '0' then
                int_reg <= "00000000";
                index <= 0;
        else
                int_reg(index) <= di;
                if index = 7 then
                        index <= 0;
                else
                        index <= index + 1;
                end if;
        end if;
    END PROCESS;
    q <= int_reg;
END simple;
```

Parallel to Serial Conversion

The parallel to serial register has two stages of operation. The first stage is to load in the parallel data. In this model, the load signal is active low and synchronous. In other words, just like the SIPO model, there is no asynchronous function and the clock is the only signal in the sensitivity list. If the load signal is high, then the data in the register is clocked out one bit at a time. Note that the Parallel to Serial Conversion (PISO) model cycles around, and does not stop after the data has been output.

```
LIBRARY ieee;
USE ieee.Std_logic_1164.ALL;
USE ieee.Std_logic_unsigned.ALL;

ENTITY piso IS
    GENERIC(n : Positive := 8); --size of register
    PORT(
            clk : IN Std_logic;
            load: IN std_logic;
            do : OUT std_logic;
             q : IN Std_logic_vector((n-1) DOWNTO 0));
END piso;

ARCHITECTURE simple OF piso IS
    SIGNAL int_reg : Std_logic_vector((n-1) DOWNTO 0);
    SIGNAL index : integer := 0;
BEGIN
    out_process : PROCESS
```

```
    BEGIN
        WAIT UNTIL rising_edge(clk);
        if load = '0' then
                int_reg <= q;
                index <= 0;
        else
                do <= int_reg(index);
                if index = 7 then
                        index <= 0;
                else
                        index <= index + 1;
                end if;
        end if;
    END PROCESS;
END simple;
```

Summary

This short chapter has demonstrated a useful function of converting serial to parallel data and vice versa. This is an extremely common task in modern Field Programmable Gate Array (FPGA) interfaces, with most communications data being in a serial format, and most processors requiring the data to be stored in parallel registers and operating in a parallel fashion.

22

ALU Functions

Introduction

A central part of microprocessors is the ALU (Arithmetic Logic Unit). This block in a processor takes a number of inputs from registers and as its name suggests carries out either logic functions (such as NOT, AND, OR and XOR) on the inputs, or arithmetic functions (addition or subtraction as a minimum). This chapter of the book will describe how the low-level logic and arithmetic functions can be implemented in VHDL.

Logic functions

If we consider a simple inverter in VHDL, this takes a single input bit, inverts it and applies this to the output bit. This simple VHDL is shown below:

```
Library ieee;
Use ieee.std_logic_1164.all;
Entity inverter is
    Port (
            A : in std_logic;
            Q : out std_logic
    );
End entity inverter;
Architecture simple of inverter is
Begin
    Q <= NOT A;
End architecture simple;
```

Clearly the inputs and output are defined as single std_logic pins, with direction in and out respectively. The logic equation is also intuitive and straightforward to implement. We can extend this be applicable to n bit logic busses by changing the entity (the architecture remains the same) and simply assigning the input and outputs the type std_logic_vector instead of std_logic as follows:

```
Library ieee;
Use ieee.std_logic_1164.all;
Entity bus_inverter is
    Port (
            A : in std_logic_vector(15 downto 0);
            Q : out std_logic_vector(15 downto 0)
    );
End entity bus_inverter;
Architecture simple of bus_inverter is
Begin
    Q <= NOT A;
End architecture simple;
```

As can be seen from the VHDL, we have defined a specific 16 bit bus in this example, and while this is generally fine for processor design with a fixed architecture, sometimes it is useful to have a more general case, with a configurable bus width. In this case we can modify the entity again to make the bus width a parameter of the model:

```
Library ieee;
Use ieee.std_logic_1164.all;
Entity n_inverter is
    Generic (
            N : natural := 16
    );
    Port (
            A : in std_logic_vector((n-1) downto 0);
            Q : out std_logic_vector((n-1) downto 0)
    );
End entity n_inverter;
Architecture simple of n_inverter is
Begin
    Q <= NOT A;
End architecture simple;
```

We can of course create separate models of this form to implement multiple logic functions, but we can also create a compact multiple function logic block by using a set of configuration pins to define which function is required. If we define a general logic block that has 2 n-bit inputs (A & B), a control bus (S) and an n-bit output (Q), then by setting the 2 bit control word (S) we can select an appropriate logic function according to the table below:

S	Function
00	Q <= NOT A
01	Q <= A AND B
10	Q <= A OR B
11	Q <= A XOR B

Clearly we could define more functions, and this would require more bits for the select function (S), but this limited set of functions demonstrates the principle involved. We can define a modified entity as shown below:

```
Library ieee;
Use ieee.std_logic_1164.all;
Entity alu_logic is
    Generic (
        N : natural:= 16
    );
    Port (
        A : in std_logic_vector((n-1) downto 0);
        B : in std_logic_vector((n-1) downto 0);
        S : in std_logic_vector(1 downto 0);
        Q : out std_logic_vector((n-1) downto 0)
    );
End entity alu_logic;
```

Now, depending on the value of the input word (S), the appropriate logic function can be selected. We can use the case statement introduced in the VHDL primer chapter of this book to define each state of S and which function will be carried out in a very compact form of VHDL:

```
Architecture basic of alu_logic is
Begin
    Case S is
        When "00" => Q <= NOT A;
        When "01" => Q <= A AND B;
        When "10" => Q <= A OR B;
        When "11" => Q <= A XOR B;
    End case;
End architecture basic;
```

Clearly this is an efficient and compact method of defining the combinatorial logic for each state of the control word (S), but great care must be taken to assign values for every combination to avoid inadvertent latches being introduced into the logic when synthesised.

1-bit adder

The arithmetic 'heart' of an ALU is the addition function – the Adder. This starts form a simple 1 bit adder and is then extended to multiple bits, to whatever sized of addition function is required in the ALU. The basic design of a 1-bit adder is to take two logic inputs (a & b) and produce a sum and carry output according to the following truth table:

a	b	sum	carry
0	0	0	0
0	1	1	0
1	0	1	0
1	1	1	1

This can be implemented using simple logic with a 2 input AND gate for the carry, and a 2 input XOR gate for the sum function as shown in Figure 63.

This function has a carry-out (carry), but no carry-in, so to extend this to multiple bit addition, we need to implement a carry-in function (cin) and a carry-out (cout) as shown in next page.

With an equivalent logic function as shown in Figure 64.

Figure 63
Simple 1-Bit
Adder

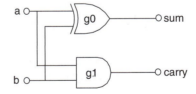

Figure 64
1-Bit Adder with
Carry-in and
Carry-out

a	b	cin	sum	Cout
0	0	0	0	0
0	1	0	1	0
1	0	0	1	0
1	1	0	0	1
0	0	1	1	0
0	1	1	0	1
1	0	1	0	1
1	1	1	1	1

This can be implemented using standard VHDL logic functions with bit inputs and outputs as follows. First define the entity with the input and output ports defined using bit types:

```
entity full_adder is
        port (sum, co : out bit;
                a, b, ci : in bit);
end entity full_adder;
```

Then the architecture can use the standard built-in logic functions in a 'dataflow' type of model, where logic equations are used to define the behaviour, without any delays implemented in the model.

```
architecture dataflow of full_adder is
begin
  sum <= a xor b xor ci;
  co <= (a and b) or
        (a and ci) or
        (b and ci);
end architecture dataflow;
```

This model is now a simple building block that we can use to create multiple bit adders structurally by linking a number of these models together.

Structural *n*-bit addition

Using the simple 1-bit full adder defined previously, it is a simple matter to create a multiple bit full adder using this model as a building block. As an example, to create a 4 bit adder, with a single carry-in and single bit carry-out, we can define a VHDL model as shown below:

```
entity four_bit_adder is
  port (sum: out bit_vector (3 downto 0); co : out bit;
        a, b : in bit_vector (3 downto 0); ci : in bit);
end entity four_bit_adder;

architecture simple of four_bit_adder is
  signal carry : bit_vector (3 downto 1);
begin
  fa0 : entity work.full_adder
     port map (sum(0),carry(1),a(0),b(0),ci);
  fa1 : entity work.full_adder
     port map (sum(1),carry(2),a(1),b(1),carry(1));
  fa2 : entity work.full_adder
     port map (sum(2),carry(3),a(2),b(2),carry(2));
  fa3 : entity work.full_adder
     port map (sum(3),co,a(3),b(3),carry(3));
end architecture simple;
```

This can obviously be extended to multiple bits by repeating the component use in the architecture for as many bits are required.

Configurable *n*-bit addition

While the structural approach is useful, it is clearly cumbersome and difficult to configure easily. A more sensible approach is to add a generic (parameter) to the model to enable the number of bits to be customised. For example, if we define an entity to add two logic vectors (as opposed to bit vectors used previously), the entity will look something like this:

```
library IEEE;
use IEEE.std_logic_1164.all;

entity add_beh is
  generic(top : natural := 15);
    port (a    : in std_logic_vector (top downto 0);
          b    : in std_logic_vector (top downto 0);
          cin  : in std_logic;
          sum  : out std_logic_vector (top downto 0);
          cout : out std_logic);
end entity add_beh;
```

As can be seen from this entity, we have a new parameter, top, which defines the size of the input vectors (a and b) and the output sum (cout). We can then use the same original logic equations that we defined for the initial 1-bit adder and use more behavioural VHDL to create a much more readable model:

```
architecture behavior of add_beh is
begin
  adder: process(a,b,cin)
          variable carry : std_logic;
          variable tempsum : std_logic_vector(top
          downto 0);
        begin
          carry := cin;
          for i in 0 to top loop
              tempsum(i) := a(i) xor b(i) xor carry;
              carry := (a(i) and b(i))
                            or (a(i) and carry)
                            or (b(i) and carry);
          end loop;
        sum <= tempsum;
        cout <= carry;
    end process adder;
end architecture behavior;
```

This architecture shows how a single process (with sensitivity list a, b, cin) is used to encapsulate the addition. The process is activated when a, b or cin changes. A for loop is used to calculated a temporary sum (tempsum) that increments each time around the loop if required and the final value is assigned to the output sum. Also, a stage by stage carry is calculated and used each time around the loop. After the final loop, the value of carry is used to become the final carry-out.

Twos complement

An integral part of subtraction in logic systems is the use of 'twos complement'. This enables us to execute a subtraction function using only an adder rather than requiring a separate subtraction function. Twos complement is an extension to the basic ones complement (or basic inversion of bits) previously considered.

If we consider an 'unsigned' number system based on 4 bits, then the range of the numbers is 0–15 (0000–1111). If we consider a 'signed' system, however, the Most Significant Bit (MSB) is considered to be the sign ($+$ or $-$) of the number system and therefore the range of numbers with 4 bits will instead be from -8 to $+7$. The method of conversion from positive to negative number in binary logic is a simple two stage process of first inverting all the bits and then adding 1 to the result.

Consider an example. Take a number 0011_2. In signed number form, the MSB is 0, so the number is positive and the lower three bits 011 can be directly translated into decimal 3. To get the twos complement (-3), we first invert all the bits to get 1100, and then add a single bit to get the final twos complement value 1101. To check that this is indeed the inverse in binary, simple add the number 0011 to its twos complement 1101 and the result should be 0000.

This function can be implemented simply in VHDL using the following model:

```
library ieee;
use ieee.std_logic_1164.all;
use ieee.numeric_std.all;

entity twoscomplement is
  generic (
      n : integer := 8
  );
  port (
      input  : in std_logic_vector((n-1) downto 0);
      output : out std_logic_vector((n-1) downto 0)
  );
end;

architecture simple of twoscomplement is
begin
  process(input)
      variable inv : unsigned((n-1) downto 0);
  begin
      inv := unsigned(NOT input);
      inv := inv + 1;
      output <= std_logic_vector(inv);
  end process;
end;
```

As can be seen from the VHDL, we operate using logic functions first (NOT) and then convert to unsigned to utilise the addition function (inv $+1$), and finally convert the result back into a std_logic_vector type. Also notice that the generic n allows this model to be configured for any data size. In this example, the test bench is used to check that the function is operating correctly by using two test circuits back to back, inverting and re-inverting the input word and checking that the function returns the same value. While this does not guarantee correct operation (the same bug could be present in both transforms!), it is a simple quick check that is very useful and makes generation of test data and checks very easy as the input and final output signal check can be XORd to check for differences:

```vhdl
library ieee;
use ieee.std_logic_1164.all;
use ieee.numeric_std.all;

entity twoscomplementtest is
end twoscomplementtest ;

architecture stimulus of twoscomplementtest is
    signal rst : std_logic := '0';
    signal clk : std_logic:= '0';
    signal count : std_logic_vector (7 downto 0);
    signal inverse : std_logic_vector (7 downto 0);
    signal check : std_logic_vector (7 downto 0);
    component twoscomplement
          port(
                  input : in std_logic_vector(7 downto 0);
                  output : out std_logic_vector(7 downto 0)
              );
    end component;
    for all : twoscomplement use entity
      work.twoscomplement ;
begin
    CUT1: twoscomplement port map(input => count,
      output => inverse);
    CUT2: twoscomplement port map(input => inverse,
      output => check);

    -- clock and reset process
    clk <= not clk after 1 us;
    process
    begin
      rst <= '0','1' after 2.5 us;
      wait;
    end process;

    -- generate data
    process(clk, rst)
      variable tempcount : unsigned(7 downto 0);
begin
  if rst = '0' then
    tempcount := (others => '0');
  elsif rising_edge(clk) then
      tempcount := tempcount + 1;
  end if;
  count <= std_logic_vector(tempcount);
  end process;
end;
```

Summary

This chapter has introduced the key elements required in an Arithmetic and Logic Unit of a processor. Whether the designer needs to implement a complete ALU from scratch, or if it is purely to understand the behaviour of an existing architecture, these functions are very useful in analysing the behaviour of ALUs and processors.

23

Decoders and Multiplexers

Decoders

Decoders are a simple combinatorial block that converts one form of digital representation into another. Usually, a decoder takes a smaller representation and converts it into a larger one (the opposite of encoding). Typical examples are the decoding of an n-bit word into 2^n individual logic signals. For example, a 3–8 decoder takes 3 logic signals in and decodes the value of one of the 8 output signals (2^3) to the selected value. The symbol for such a decoder is given in Figure 65 with its functional behavior in the following table:

s2	s1	s0	q7	q6	q5	q4	q3	q2	q1	q0
0	0	0	0	0	0	0	0	0	0	1
0	0	1	0	0	0	0	0	0	1	0
0	1	0	0	0	0	0	0	1	0	0
0	1	1	0	0	0	0	1	0	0	0
1	0	0	0	0	0	1	0	0	0	0
1	0	1	0	0	1	0	0	0	0	0
1	1	0	0	1	0	0	0	0	0	0
1	1	1	1	0	0	0	0	0	0	0

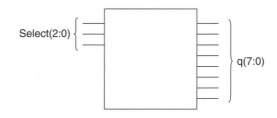

Figure 65
3–8 Decoder

The VHDL for this decoder uses a simple VHDL construct similar to the if – else – end if form, except using the when – else syntax. If a signal is assigned a value when a condition is satisfied, then a single assignment can be made using the following basic pseudo-code:

```
output <= value when condition;
```

This can be extended with else statements to cover a set of different conditions, thus:

```
output <=      value1 when condition1 else
               value2 when condition2 else
               ...
               valuen when condition;
```

Finally, if there is a 'catch all' condition, similar to the final else in an if – elsif – else – endif conditional statement in VHDL, then the final assignment would be added as follows:

```
output <=      value1 when condition1 else
               value2 when condition2 else
               ...
               valuen when conditionn else
               valuedefault;
```

Using this approach, the 3–8 decoder can be simply implemented using the following VHDL:

```
library ieee;
use ieee.std_logic_1164.all;
use ieee.numeric_std.all;

entity decoder38 is
  port (
        s : in std_logic_vector (2 downto 0);
        q : out std_logic_vector(7 downto 0)
  );
end;

architecture simple of decoder38 is
begin
        q <= "00000001" when s = "000" else
        "00000010" when s = "001" else
        "00000100" when s = "010" else
        "00001000" when s = "011" else
        "00010000" when s = "100" else
        "00100000" when s = "101" else
        "01000000" when s = "110" else
        "10000000" when s = "111" else
        "XXXXXXXX";
    end;
```

The test bench for this decoder could be a simple look-up table of values, but in fact we could combine the clock and reset test bench from the counter example, and include a simple counter in the test bench to generate the signals input to the decoder as follows:

```vhdl
library ieee;
use ieee.std_logic_1164.all;
use ieee.numeric_std.all;

entity Decoder38Test is
end Decoder38Test;

architecture stimulus of Decoder38Test is
      signal rst : std_logic := '0';
      signal clk : std_logic:='0';
      signal s : std_logic_vector(2 downto 0);
      signal q : std_logic_vector(7 downto 0);

      component decoder38
          port(
                  s : in std_logic_vector(2 downto 0);
                  q : out std_logic_vector(7 downto 0)
              );
      end component;
      for all : decoder38 use entity work.decoder38;
begin

      CUT: decoder38 port map(s => s, q => q);
      clk <= not clk after 1 us;
      process
      begin
       rst <= '0','1' after 2.5 us;
       wait;
      end process;

process(clk, rst)
      variable count : unsigned(2 downto 0);
begin
      if rst = '0' then
         count := (others => '0');
      elsif rising_edge(clk) then
         count := count + 1;
      end if;
      s <= std_logic_vector(count);
end process;

end;
```

Multiplexers

A multiplexer is an extension of a simple decoder in that a series of inputs are decoded to provide select enables for one of a number

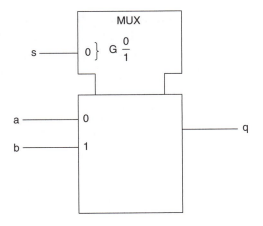

Figure 66
2 Input Multiplexer with
a single select line

of inputs. In a similar way that *n*-bits can decode 2^n signals, in a multiplexer, *n*-bits of select line are required to multiplex 2^n signals. For example, consider the simplest multiplexer, a 2 input (A and B), single output (Q) multiplexer, with a single select line (S). The IEEE symbol for such a MUX is given in Figure 66.

A similar approach to the decoder by using the when – else structure can be used to create a simple implementation of the multiplexer as shown in the following VHDL:

```
library ieee;
use ieee.std_logic_1164.all;
use ieee.numeric_std.all;

entity mux21 is
  port (
        s : in std_logic;
        a : in std_logic;
        b : in std_logic;
        q : out std_logic
  );
end;

architecture simple of mux21 is
begin
        q <= a when s = '0' else
        b when s = '1' else
        'X';
end;
```

This is an extremely useful model and is extensively used in test structures where it is required to choose between a functional and test input signal input to a flip-flop. The model can be easily extended to accommodate multiple input signals. For example, consider a 4 input multiplexer, with 2 select signals (inputs = 2select) and a single output. The VHDL model has largely the same structure, but would look like this:

```vhdl
library ieee;
use ieee.std_logic_1164.all;
use ieee.numeric_std.all;

entity mux21 is
  port (
        s : in std_logic_vector (1 downto 0);
        a : in std_logic;
        b : in std_logic;
        c : in std_logic;
        d : in std_logic;
        q : out std_logic
  );
end;

architecture simple of mux21 is
begin
        q <= a when s = "00" else
        b when s = "01" else
        c when s = "10" else
        d when s = "11" else
        'X';
end;
```

Summary

This short chapter has describe the basic mechanism for decoding and multiplexing signals using VHDL. This is an extremely useful function as it is central to much of the data and control signal management on Field Programmable Gate Arrays (FPGAs).

24

Finite State Machines in VHDL

Introduction

Finite State Machines (FSM) are at the heart of most digital design. The basic idea of a FSM is to store a sequence of different unique states and transition between them depending on the values of the inputs and the current state of the machine. The FSM can be of two types Moore (where the output of the state machine is purely dependent on the state variables) and Mealy (where the output can depend on the current state variable values AND the input values). The general structure of an FSM is shown in Figure 67.

State transition diagrams

The method of describing FSM from a design point of view is using a state transition diagram (bubble chart) which shows the

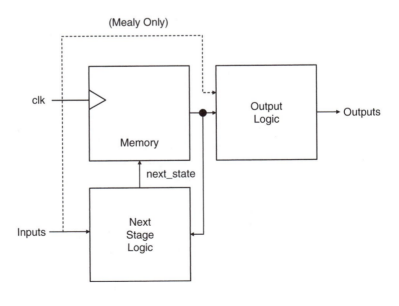

Figure 67
Hardware State
Machine Structure

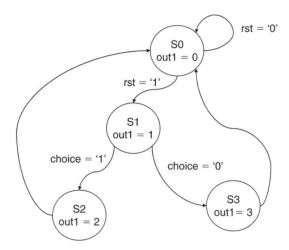

Figure 68
State Transition
Diagram

states, outputs and transition conditions. A simple state transition diagram is shown in Figure 68.

Interpreting this state transition diagram it is clear that there are four bubbles (states). The transitions are controlled by two signals ('rst' and 'choice'), both of which could be represented by bit or std_logic types (or another similar logic type). There is an implicit clock signal, which we shall call 'clk' and the single output 'out1'.

Implementing FSM in VHDL

This transition diagram can be implemented using a case statement in a process using the following VHDL:

```
library ieee;
use ieee.std_logic_1164.all;

entity fsm is
  port (
      clk, rst, choice : in std_logic;
      count : out std_logic
  );
end entity fsm;
architecture simple of fsm1 is
  type state_type is ( s0, s1, s2, s3 );
  signal current, next_state : state_type;
begin
  process ( clk )
  begin
      if ( clk = '1' ) then
            current <= next_state;
      end if;
  end process;
```

```
process (current)
begin
    case current is
        when s0 =>
            out <= 0;
            if (rst = '1') then
                next <= s1;
            else
                next <= s0;
            end if;
        when s1 =>
            out <= 1;
            if (choice = '1') then
                next <= s3;
            else
                next <= s2;
            end if;
        when s2 =>
            out <= 2;
            next <= s0;
        when s3 =>
            out <= 3;
            next <= s0;
    end case;
end process;
end;
```

Summary

FSM are a fundamental technique for designing control algorithms in digital hardware. This chapter of this book is purely an introduction to the key concepts and if the reader is not already fully familiar with the basic concepts of digital hardware design, you are encouraged to obtain a standard text on digital design techniques to complement the practical implementation methods described in this book.

25

Fixed Point Arithmetic in VHDL

Introduction

In VHDL we have complete access to a range of types from bits and Booleans (which consist of two states '0' and '1' (or true and false) which are effectively enumerated types, through integer numbers (including positive and natural subtypes) and eventually we can use real numbers (floating point). Unfortunately, the big drawback is not necessarily what we can use in VHDL, but rather what we can synthesize in hardware.

Despite recent research efforts and standardization efforts, there is still a limited availability of packages and libraries that support both fixed point and floating point arithmetic specifically for Field Programmable Gate Arrays (FPGAs). If we consider most FPGA applications, there is a need for some DSP type application, and generally a form of fixed point arithmetic will be adequate in most of these cases.

So, what is fixed point arithmetic and how can we use it in FPGA design? In integer arithmetic, unsigned, signed or std_logic, the basis of the number is a bitwise representation of an integer number, with no decimal point. For example, to represent the number 23, using 8 bits, we simply set a bit for each binary element required to construct the integer value of 23. This is shown in Figure 69.

128	64	32	16	8	4	2	1
0	0	0	1	0	1	1	1

Figure 69
Basic Binary
Notation

$$16 + 4 + 2 + 1 = 23$$

If we require a negative number, then we use the 'signed' approach, where the Most Significant Bit (MSB) is simply the sign bit as shown in Figure 70. In fact, the two's complement notation (discussed previously in the chapter on Arithmetic Logic Unit (ALU) functions), can be obtained by inverting the bits and adding one to the LSB.

With this basic idea of handling numbers, we can extend the notation to a 'fixed point' scheme by defining where the decimal point will go. For example, in the same number scheme shown we have 8 bits. We can therefore define this in terms of 5 bits above the decimal point and 3 below it. This will give some limited fractional usage for the numbers. The way that this is implemented is by using fractions of 1 for each 'negative' bit to the right of the decimal point. As an example, take the same number in terms of bits used in Figure 70 and use the new fixed point numbering system for the bits. In this case we get a value of -2.875 (Figure 71).

The nice thing about this notation is that the bitwise functions defined for the integer-based ALU developed previously can also be applied to this new fixed point notation with almost no modification. The only difference is that we need to translate from the new fixed point type toa std_logic_vector type and also consider how to handle overflow conditions.

For example, if two numbers are added together and the result is too large, how is this handled by the fixed point algorithms? Do we simply flag an overflow and output the result? Or do we set the maximum value and output this?

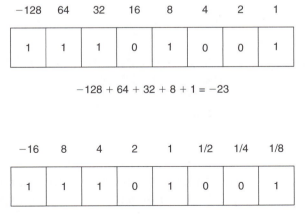

Figure 70
Negative Number
Binary Notation

Figure 71
Fixed Point Notation

Similarly, for numbers which may be too small, and we can potentially lose precision, do we simply round up or flag another loss of precision condition? These are questions that the designer needs to answer for their application, but for the rest of the chapter a simple approach will be taken that illustrates how the basic functions operate, and the details of handling these issues will be left to the reader, unless specifically identified and discussed.

Basic fixed point types

The first task in defining a custom fixed point library, is to specify a new type for the numbers. The closest similar types in standard VHDL, that can be synthesized, are unsigned and signed. These are defined in terms of a specific number of bits. In most cases we are interested in linking directly to std_logic systems, and so in this case we can effectively define a new type based on an array of std_logic bits. For the remainder of this chapter we will discuss signed arithmetic only, as this is the most potentially used from a DSP and application point of view.

The basic type that defines our base type is to be called fixsign and is defined as an unrestricted array of std_logic:

```
Type fixsign is array ( integer range <> ) of std_logic;
```

From this, we can define specific subtypes that have a defined range of fixed point. For example, we can define a type that has 8 bits above the decimal point and 3 bits below using the following declaration:

```
Subtype fp8_3 is fixsign ( 8 downto -3);
```

Using these new types we can declare signals of this new type and use it in fixed point VHDL models:

```
Signal a1 : fp8_3;
A1 <= X"0CA";
```

Clearly this is useful, but limited as this type needs to be able to be converted from one type to another easily and quickly. The simplest way to manage this process is to create a new package that contains not only the type declarations, but also the functions that are associated with this set of types. Therefore we can define a new package called fp_pkg that as a minimum contains these type declarations:

```
package fp_pkg is
    type fixsign is array (integer range <>) of std_logic;
    subtype fp8_3 is fixsign ( 8 downto -3);
end package;

package body fp_pkg is
end package body;
```

We can now use this package in a VHDL model by compiling the package into the current work library and calling the package as we need it:

```
Use work.fp_pkg.all;
```

This will provide access to all the fixed point functions and types required.

In this library, we have two types of functions. The first type are required for translating physical types (such as std_logic_vector) to our new types and vice versa. These are important as they will be synthesized and eventually end up on hardware. The second type are purely for debug purposes and displaying values to the screen. For example, it is useful to be able to convert fixed point data to real numbers and then use the real'image VHDL function to display the value to the screen. A useful set of functions is therefore presented in this chapter. Again, these are exemplar functions, and the reader is encouraged to develop these basic functions and produce their own for their own applications.

Fixed point functions

Fixed point to std_logic_vector functions

The most important functions are the conversion between fixed point and std_logic_vector variables. If we can translate from one to the other, then we can use our standard logic functional blocks where appropriate on the fixed point data directly, rather than needing to come up with brand new blocks every time.

The easiest function is the mapping from fixed point to std_logic_vector and is simply a matter of starting from the LSB defined in the range of the fixed point number and then setting each bit on the output std_logic_vector in turn to the correct value. The VHDL for this is given below:

```
function fp2std_logic_vector (d:fixsign;top:integer;
   low:integer)
      return std_logic_vector is
      variable outval : std_logic_vector ( top-low
        downto 0 ) := (others => '0');
begin
      for i in 0 to top-low loop
             outval(i) := d(i + low);
      end loop;
   return outval;
end;
```

If we look at this function we can see that the arguments to the function are the fixed point number, and then the two integer values that denote the number of bits above and below the decimal point, respectively. For example, if our notation is 8.3, the function call in this case would be:

```
Q <= fp2std_logic_vector(d,8,-3);
```

Notice the negative number denoting the bits below the decimal point. If you would prefer both numbers to be positive, they can simply be changed. One reason for using the negative form, is that the numbers match the basic type definition and therefore make checking easy.

Similarly, we can convert from std_logic_vector back to fixed point using a very similar function in the opposite direction:

```
function std_logic_vector2fp
   (d:std_logic_vector;top:integer;low:integer)
return fixsign is
      variable outval : fixsign ( top downto low )
        := (others => '0');
begin
      for i in 0 to top-low loop
             outval(i + low) := d(i);
      end loop;
      return outval;
end;
```

With the similar usage:

```
Q <= std_logic_vector(d,8,-3);
```

Using these functions, the conversion between the std_logic_vector and fixed point arithmetic domains becomes straightforward. Also, these functions are easily synthesizable as they simply map bits and do not carry out any sophisticated functions other than that.

Fixed point to real conversion

An extremely useful function is the ability to convert from fixed point to real numbers. Obviously this is no use for synthesis, but is ideal for adding, checking and reports to test benches. As a result we only define a single function fp2real which takes a fixed point number and converts in to a real number for display. Once we have the number, then the real'image function can be used to display the value. The VHDL for the conversion function is given below:

```vhdl
function fp2real (d:fixsign; top:integer; low:integer)
return real is
        variable outreal : real := 0.0;
        variable mult : real := 1.0;
        variable max : real := 1.0;
        variable debug : boolean := false;
begin
        for i in 0 to top-1 loop
            if d(i) = '1' then
                    outreal := outreal + mult;
                    if debug  then
                       report "  fp2real : " &
                          integer'image(i);
                    end if;
            end if;
            mult := mult * 2.0;
        end loop;
if debug  then
   REPORT " fp2real middle : " & real'image(outreal);
end if;
max := mult;

mult := 0.5;

for i in -1 downto low loop
    if d(i) = '1' then
            outreal := outreal + mult;
            if debug  then
               report "  fp2real : " & integer'image(i);
            end if;
    end if;
    mult := mult * 0.5;
end loop;
```

```
if debug then
  REPORT " fp2real : " & real'image(outreal);
end if;

if d(top) = '1' then
      outreal := outreal - max;
end if;
if debug then
  REPORT " fp2real FINAL VALUE : " &
      real'image(outreal);
end if;

return outreal;
end;
```

This function is a simple converter that handles the bits above and below the decimal point in turn. Also notice the internal Boolean debug variable that allows checking of each individual bit. This can be very useful when observing the passing of numbers across boundaries ensuring correct translation – this defaults to false (off).

If we need to report a fixed point value, we can therefore use this function to report the values using simple VHDL such as this:

```
D : fp8_3;
Dr : real;
Dr <= fp2real(fp8_3,8,-3);
Report "The value is : " & real'image(Dr);
```

Testing the fixed point function

As stated previously, we can use these functions to incorporate standard std_logic ALU functions into the model. In this simple test case, we are using the standard n-bit adder created in the ALU functions chapter of this book to add two fixed point numbers together. How does this work? What we do is convert the two input fixed point numbers into std_logic_vectors, apply them to the adder block, then convert the output back to a fixed point number. We can convert both inputs and outputs into real numbers for observation on the screen:

```
library ieee;
use ieee.std_logic_1164.all;
use ieee.numeric_std.all;

use work.fp_pkg.all;
```

```
entity simple1 is
end entity simple1;

architecture tb of simple1 is
        signal clk : std_logic := '0';
        signal cin : std_logic := '0';
        signal cout : std_logic;
        signal testa : fp8_3 := "000000000000";
        signal testa1 : fixsign ( 8 downto -3 );
        signal testa2 : fixsign ( 8 downto -3 );
        signal testb1 : fixsign ( 8 downto -3 );
        signal testsum : fixsign ( 8 downto -3 );
        signal as : signed ( 11 downto 0) := X"000";
        signal a1std : std_logic_vector ( 11 downto 0) :=
          X"800";
        signal b1std : std_logic_vector ( 11 downto 0) :=
          X"800";
        signal sum : std_logic_vector ( 11 downto 0) ;
        signal a1out : real;
        signal b1out : real;
        signal a2out : real;
        signal sumout : real;
        signal a1 : integer := 0;
        signal bs : signed ( 11 downto 0) := X"8f0";

        component add_beh
        generic (
                top : integer := 7
        );
        port (
                signal a : in std_logic_vector(top
                  downto 0);
                signal b : in std_logic_vector(top
                  downto 0);
                signal cin : in std_logic;
                signal cout : out std_logic;
                signal sum : out std_logic_vector(top
                  downto 0)
        );
        end component;
        for all : add_beh use entity work.add_beh;

begin
        clk <= not clk after 1 us;

        DUT :add_beh generic map ( 11 ) port map ( a1std,
          b1std, cin, cout, sum);

        p1 : process (clk)
        begin
                as <= as + 1;
                testa1 <= signed2fp(as,8,-3);
                testb1 <= signed2fp(bs,8,-3);
```

```
                              a1out <= fp2real(testa1,8,-3);
                              b1out <= fp2real(testb1,8,-3);
                              a1std <= fp2std_logic_vector(testa1,8,-3);

                              b1std <= fp2std_logic_vector(testb1,8,-3);
                              testa2 <= std_logic_vector2fp(a1std,8,-3);
                              testsum <= std_logic_vector2fp(sum,8,-3);
                              a2out <= fp2real(testa2,8,-3);
                              sumout <= fp2real(testsum,8,-3);
                              report "a1out : " & real'image(a1out);
                              report "a2out : " & real'image(b1out);
                              report "sumout : " & real'image(sumout);

                      end process p1;
              end;
```

An important aspect to note in this model is the use of signals and a clock (clk). By making this model synchronous, we have ensured correct, predictable behavior, but on each clock cycle there are several delays built in. The final observed result on sumout (the real number output for display) will appear 2 clock cycles after the data is input to the model.

In this case we are using signed numbers as the original input (as) as these can be incremented easily and setting one number to a constant (bs). These inputs are converted to real numbers (a1out, b1out) that are displayed to the screen to show the results.

Summary

This chapter has introduced the concept of fixed point arithmetic in VHDL and provided a basic package of functions and types to get started using VHDL. It must be stressed that this package is purely for exemplar designs and the reader is encouraged to either use commercially available libraries for optimum performance or to develop their own libraries.

26

Binary Multiplication

Introduction

A key function in any hardware design that requires signal processing is multiplication. In order to implement such a function it is useful to introduce the basic methods for binary multiplication from first principles so that the implemented approaches can be understood. In this chapter we will describe these methods and illustrate them with VHDL.

Basic binary multiplication

The simplest approach to binary multiplication is essentially long multiplication applied to binary numbers. Consider a basic example of a decimal long multiplication first to remind us of the basic concept – take a multiplication of two numbers 23 and 17:

$$
\begin{array}{r}
23 \\
\times\,17 \\
\hline
1\;6\;1 \\
2\;3\;0 \\
\hline
3\;9\;1 \\
\hline
\end{array}
$$

This can be implemented using binary numbers in exactly the same way, except instead of decimal numbers, the arithmetic is binary. Consider the multiplication of two unsigned binary numbers for 6 (0110) and 4 (0100). Simply take each bit of the multiplier (4 in this case) and if it is zero, add nothing, but if the bit is one, add the shifted multiplicand (6 in this case).

$$
\begin{array}{rl}
0110 & (6) \\
\times\ 0100 & (4) \\
\hline
0000 & \\
0000 & \\
0110 & \\
0000 & \\
\hline
011000 & (24)
\end{array}
$$

The way we implement this in practice is to have a 'partial product' and then add the shifted multiplicand (or zeros) at each stage of the process until the multiplication is complete.

While this approach works for unsigned binary numbers, it does not work for twos complement numbers. In the case of twos complement, using a similar approach requires the addition of sign bits to the left of the shifted multiplicand at each stage and then a final step of negating the multiplicand and adding the final shifted value to the partial product. A simpler approach that lends itself well to hardware implementation is simply to test whether a number (or both) are negative, invert to obtain the magnitude of each number if necessary, carry out an unsigned multiplication, then depending on how many of the arguments are negative – invert the output (twos complement). The method of checking for negative numbers is relatively simple, as an XOR function on the Most Significant Bit (MSB) of the two input signed numbers will tag whether the output needs to have a twos complement taken. This is shown schematically in Figure 72.

VHDL unsigned multiplier

If we start with a simple unsigned multiplier, then this can be implemented very simply using VHDL. The important aspect to consider with this multiplier is how many bits will be on the inputs and how many on the outputs. If the number of bits are the same across all three, then we need to consider the possibility of overflow and how this can be handled by the multiplier. In this basic model, we will define the output number of bits as being the sum of the two input word lengths, and deal with overflow externally to the multiplier.

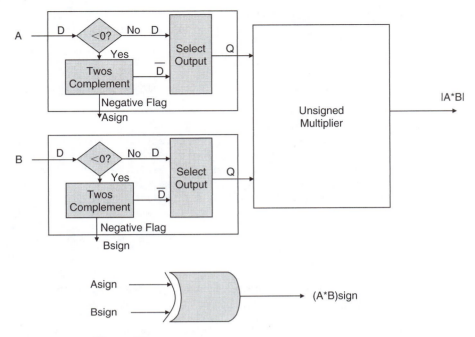

Figure 72
Basic Signed Multiplication

If we use the basic accumulator and addition function of the simple binary addition method described previously, we can implement a basic VHDL multiplier as shown below:

```
library ieee;
use IEEE.std_logic_1164.all;

entity mult_beh is
  generic(top : natural := 15);
    port (
        clk  : in std_logic;
        nrst : in std_logic;
          a : in std_logic_vector (top downto 0);
          b : in std_logic_vector (top downto 0);
          product : out std_logic_vector (2*top+1
            downto 0)
    );
end entity mult_beh;

architecture behavior of mult_beh is
      component add_beh
      generic (
            top : integer := 7
      );
      port (
            signal a : in std_logic_vector(top downto 0);
            signal b : in std_logic_vector(top downto 0);
            signal cin : in std_logic;
```

```vhdl
                    signal cout : out std_logic;
                    signal sum : out std_logic_vector
                    (top downto 0)
            );
        end component;
        for all : add_beh use entity work.add_beh;

        signal cin : std_logic := '0';
        signal cout : std_logic := '0';
        signal acc : std_logic_vector(2*top+1 downto 0);
        signal sum : std_logic_vector(2*top+1 downto 0);
        signal mand : std_logic_vector(2*top+1 downto 0);
        signal index : integer := 0;
        signal finished : std_logic := '0';
begin

        DUT :add_beh generic map (2*top+1) port map
           (acc,mand,cin,cout,sum);

    p1 : process (clk, nrst)
        variable mandvar : std_logic_vector(2*top+1 downto 0);
    begin
            if (nrst = '0') then
                acc <= (others => '0');
                index <= 0;
                finished <= '0';
            else
                if rising_edge(clk) then
                    if index <= top then
                            index <= index [Plus] 1;
                            mandvar := (others => '0');
                            if b(index) = '1' then
                              for i in 0 to top loop
                                 mandvar(i+index) := a(i);
                              end loop;
                            end if;
                    end if;
                    mand <= mandvar;
                    acc <= sum;
                end if;
                if falling_edge(clk) then
                    if index > top-1 then
                            finished <= '1';
                    end if;
                end if;
            end if;
    end process p1;
    p2 : process (finished)
    begin
            if rising_edge(finished) then
                product <= sum;
            end if;
    end process p2;
end architecture behavior;
```

This model is perhaps more complex than it really needs to be, but it does have some nice features from a learning point of view.

Firstly, rather than a 'super efficient' shifting model which is difficult to read, the shift and add function in process p1 is laid out in detail so each stage of the multiplication can be followed through. Also notice the use of the signal finished which is used to show when the calculation is completed. This is useful when designing a controller to show that the calculation has been completed.

Synthesis of the multiplication function

After completion, this model was run through a standard synthesis software tool, targeted at a reasonable sized Virtex II Pro FPGA with the following results:

```
Number of ports :                        66
Number of nets :                       1704
Number of instances :                  1639
Number of references to this view :   0
Total accumulated area :
Number of BUFGP :                         1
Number of Dffs or Latches :            164
Number of Function Generators :       1181
Number of IBUF :                         33
Number of MUX CARRYs :                   31
Number of MUXF5 :                       221
Number of MUXF6 :                         2
Number of OBUF :                         32
Number of accumulated instances : 1701
Number of global buffers used:            1
***************************************************
Device Utilization for 2VP2fg256
***************************************************
Resource              Used    Avail    Utilization
--------------------------------------------------
IOs                    65      140       46.43%
Global Buffers          1       16        6.25%
Function Generators  1181     2816       41.94%
CLB Slices            591     1408       41.97%
Dffs or Latches       164     3236        5.07%
Block RAMs              0       12        0.00%
Block Multipliers       0       12        0.00%
--------------------------------------------------
    Clock              : Frequency
    -------------------------------------------
    clk                : 30.0 MHz
    finished           : 30.0 MHz
```

What is clear from this report is the fact that a significant amount of resources were required to implement this multiplier on a standard device. In this case, the optimization was for area and not speed, but in spite of that, the design usage was nearly 50 per cent of the whole FPGA (Field Programmable Gate Array), so clearly arithmetic functions are not always easy on an FPGA, certainly not in area terms, with the worst culprit being multipliers.

As a result, care must be taken in managing designs, taking advantage of pipelining and using the available resources as effectively as possible. The downside is that the design becomes more involved, with a controller generally required, but ultimately the possibility of higher performance than an equivalent DSP function.

'Simple' multiplication

As we have seen in the previous example, there is a method of implementing multiplication operations using a 'first principles' approach and it is incredibly hungry in terms of both resources and time (taking n shifts to complete a multiplication would lead to a really slow device).

There is, however, an alternative approach with many modern FPGAs that include multiplier blocks as part of the design. These are custom multiplication blocks already in place on the FPGA and this allows the specific multiply function to be implemented directly in the VHDL.

We can therefore convert the std_logic_vector signals into signed signals and then apply the product equation directly using the following VHDL (remember a and b are the two inputs, both of type std_logic_vector, and product is the output, also of type std_logic_vector).

```
Product <= std_logic_vector( signed(a) * signed(b) );
```

Clearly this is much more efficient VHDL than the previous model, but also remember to declare the IEEE numeric standard library:

```
Use ieee.numeric_std.all;
```

This allows the use of the signed variable types. The complete model using this approach is much more compact and is shown below:

```
library ieee;
use IEEE.std_logic_1164.all;
use ieee.numeric_std.all;
```

```
entity mult_sign is
  generic(top : natural := 15);
    port (
        clk : in std_logic;
        nrst : in std_logic;
        a : in std_logic_vector (top downto 0);
         b : in std_logic_vector (top downto 0);
          product : out std_logic_vector
            (2*top+1 downto 0)
    );
end entity mult_sign;

architecture behavior of mult_sign is
begin
      p1 : process (a,b)
      begin
            product <= std_logic_vector(signed(a) *
              signed(b));
          end process p1;
end architecture behavior;
```

The resulting synthesis output is much more compact. Clearly the number of IO Blocks (IOBs) will remain the same, but the usage internally on the FPGA will be much reduced:

```
Number of ports :                          66
Number of nets :                           128
Number of instances :                      65
Number of references to this view :         0
Total accumulated area :
 Number of Block Multipliers :              1
 Number of gates :                          0
 Number of accumulated instances :         65
Number of global buffers used:              0
**************************************************
Device Utilization for 2VP2fg256
**************************************************
Resource            Used    Avail    Utilization
------------------------------------------------
IOs                  66      140      47.14%
Global Buffers       0       16       0.00%
Function Generators  0       2816     0.00%
CLB Slices           0       1408     0.00%
Dffs or Latches      0       3236     0.00%
Block RAMs           0       12       0.00%
Block Multipliers    1       12       8.33%
```

Clearly, for this device, there are 12 multipliers available, and we have used only one, so the utilization of the remainder of the device is zero. This does lead to the ability to implement certain lower order filters very effectively using devices such as these.

Summary

This chapter has introduced some techniques for implementing multiplication in VHDL for FPGAs and has highlighted the clear difference between using a 'first principles' approach as opposed to utilizing the available resources on the FPGA, both in terms of area usage, and also in terms of model complexity.

There are, of course, other topologies of multiplier, including the Booth multiplier to name but one, and these are commonly used in hardware. The reader is encouraged to investigate different options for implementing hardware such as multipliers and how best to implement the function for their own application.

27
Bibliography

Introduction

It is normal in a book such as this to have a bibliography that simply lists a series of books, however in this book I have decided to not only list the book titles and details, but also give my perspective on their applicability and context to help the reader in deciding which would be a suitable book for them. Of course, this is limited to my own viewpoint and others may well disagree with my short synopses of the books, but hopefully it will help the reader understand where I found each book useful in this work.

Useful texts for VHDL

Digital Systems Design

Digital System Design with VHDL by Mark Zwolinski, published by Pearson Education is a superb introduction to designing with VHDL. It is used in many Universities worldwide for teaching VHDL at an undergraduate level and has numerous basic examples to enable a student to get started. I would also recommend this to an FPGA engineer getting started with VHDL.

Designers Guide to VHDL

The Designers Guide to VHDL by Peter Ashenden is perhaps the most comprehensive book on VHDL from a variety of perspectives. It covers the syntax and language rigorously, has plenty of examples, and is a great desk top reference book. For non-beginners in VHDL, this is the book I would recommend.

VHDL: Analysis and Modeling of Digital Systems

VHDL: Analysis and Modeling of Digital Systems (McGraw-Hill Series in Electrical and Computer Engineering) by Zainalabedin Navabi is a detailed look at not only how VHDL can be used to model digital systems, but many of the detailed issues regarding timing and analysis that are often skipped over by other texts on VHDL. It is perhaps not a beginners book, but is especially useful for those who require a deeper understanding of issues relating to timing.

VHDL for Logic Synthesis

VHDL for Logic Synthesis by Andrew Rushton, published by Wiley, is a useful background text for those who perhaps need to understand how VHDL can be used for practical synthesis. The book discusses what and what is not synthesizable and also explains how some useful and somewhat arcane VHDL functions operate.

Useful Texts for FPGAs

Design Warriors Guide to FPGAs

A Design Warriors Guide to FPGAs by Clive 'Max' Maxfield, published by Elsevier, is an excellent introduction to the field of FPGAs. It introduces the main concepts in designing with FPGAs as the platform and does not get into low level details of VHDL or Verilog, but does have a balance between high level design issues and low level details. This is especially useful for the student who needs to know how FPGAs work and also for engineers who need a 'heads up' on how FPGAs can be used in practice.

General Digital Design Books

Digital Design

Digital Design, by M. Morris Mano, published by Prentice Hall, is a good background text for digital design and computer design. A particularly useful aspect for those designing embedded processors is the section of the book that discusses the difference between high level languages, assembly language and machine code and then develops that into a design methodology. For anyone starting out with processor design this is a very useful text. Mano also has a related book called *Computer System Architecture* that has more detail in this area and is equally useful.

Index